OPEN LIBRARY SCIENCE

RELATIVITY

OPEN LIBRARY ◎ SCIENCE

General editor Frank Barnaby

RELATIVITY

An elementary, non-mathematical introduction to classical, special and general relativity

John Marks

Illustrations by Sandra Barnaby

GEOFFREY CHAPMAN
LONDON 1972

Geoffrey Chapman Publishers
35 Red Lion Square, London WC1 4SJ

Geoffrey Chapman (Ireland) Publishers
5-7 Main Street, Blackrock, County Dublin

ISBN 0 225.65857.7

First published 1972

This book is set in 10 on 12pt, Times Roman and printed in Great Britain by A. Wheaton & Company, Exeter.

Contents

Preface

The usual meaning of the word 'relativity' is something like 'the quality or state of being relative', to quote a definition from a well-known dictionary. To the scientist, however, 'relativity' means 'the study of how the laws of physics appear to different observers who are moving with respect to one another' and this book is concerned with this more restricted meaning of the word. Relativity, in the scientific sense of the word, therefore deals with the relative motion of two or more observers of the physical world, and in particular with how this relative motion affects the measurements made by the observers and the laws they deduce from these measurements.

For many years this subject was not considered to be particularly interesting or significant since the effects of relative motion were thought to be well understood on the basis of a few simple common-sense ideas. These ideas were usually implicitly incorporated into scientific theories as they were needed and the term relativity was seldom used.

Then, in the early years of this century, Albert Einstein took a fresh and searching look at these accepted ideas and found that they involved some inconsistencies and illogicalities. After considerable effort, he was able to reconstruct relativity in a more satisfactory way but he then found that his new ideas led to some surprising predictions. In particular, they predicted such unexpected things as that time was not absolute but different for different observers, that the dimensions and masses of material objects are not fixed but change in unexpected ways when the objects are moving, and, perhaps most astonishing of all, that mass is really nothing but a very concentrated form of energy. At the time there was no direct evidence for any of these predictions and Einstein was an unknown patents clerk of 26. So his ideas were not well received. Gradually, however, evidence for his ideas accumulated and their very revolutionary nature began to capture the imaginations of scientists and also of the general public.

This book explains his ideas to the general reader who has no previous knowledge of the subject and shows how these surprising conclusions stem from a few basic principles. It is not a mere catalogue of the results of Einstein's theories but an attempt to present the theories in a serious but elementary way and to give the reader some feeling for the subject and its development. Consequently it devotes quite a lot of space to the background of science in general and of physics in particular, and also tries to put Einstein's ideas in a wider perspective. The book is non-mathematical and makes extensive use of graphical methods to illustrate quantitative ideas.

To sum up, the book tries to be comprehensive without being superficial or incomprehensible and should therefore be useful to readers with some command of mathematics and physics and to serious students of relativity as well as to the general reader.

John Marks

1

Introduction and Some Essential Preliminaries

Most people have heard of Einstein, and many people who are interested in scientific matters know that he did something extraordinary—but few even among scientists have a really clear picture of his most famous achievements, the theories of special and general relativity. This is probably because his work is usually presented in isolation and, even in popular accounts, a considerable amount of mathematical detail is included. Einstein's ideas, however, do not stand in isolation; they are part of the continuous development of physical science from the time of Galileo to the present day. And while the complete exposition of his work is necessarily mathematical, many of his ideas may be appreciated without the use of algebra or other symbolic forms of mathematical expression.

This book attempts to discuss Einstein's work seriously but in a non-mathematical way and to present his ideas in the context of preceding and subsequent work on the nature of the physical world. Therefore the early chapters are devoted to explaining some basic ideas about science and how it progresses, and to presenting the ideas on the physical world which were generally accepted at the beginning of the twentieth century before Einstein's first paper on special relativity was published in 1905. Later chapters deal with the ideas of special and general relativity and discuss the main experimental evidence which is relevant to the theories. The final chapter ranges more widely and touches on some of the more controversial aspects of relativity theory including its present and future status.

The rest of this chapter will deal with some common difficulties which non-scientists experience when attempting to understand scientific work. Probably the main general difficulties which are encountered are the use of everyday words with special technical meanings, the quantitative nature of the physical sciences which leads to the widespread use of mathematics, and, particularly in the case of relativity theory, the prediction of effects which seem to fly in the face of common sense. These difficulties will be dealt with in turn.

The first difficulty usually concerns words such as force or energy which are familiar everyday words that have been taken over into the language of science with particular qualitative and quantitative meanings. Such meanings are often akin to the common usage but obviously not exactly equivalent to it. These words will be carefully defined when they are first used in the stricter scientific sense and the words themselves will be printed in italics for the first three or four times that they are used in their exact sense. A particular problem in relativity is that even such words as time or distance fall into this category, as we shall see in later chapters. Technical terms which are not current in everyday language will also be carefully defined but will only be printed in italics the first time that they are used.

The second difficulty arises because of the major part played by precise quantitative measurements in physical science. Many physical measurements can be made so accurately that physical theories which attempt to correlate these measurements must themselves be capable of precise quantitative predictions. This inevitably means that physical theories must be expressed in mathematical terms and no serious account of such theories can be given unless some quantitative relationships are used. In this book quantitative relationships will be expressed in graphical or diagrammatical form and, where appropriate, very simple numerical examples will be used. In this way the usual mental blockage which results at the sight of a mathematical formula (in some scientists as well as non-scientists!) will, I hope, be avoided. However, some of the ideas involved are difficult and will require careful thought, whether or not they are expressed in algebraic formulae or other mathematical symbols. The main quantitative ideas which are needed in the rest of the book will now be developed with the use of graphical means.

A number of useful ideas can be obtained from Fig. 1.1, which shows graphically the relationship between the height and age of a typical human being. The graph is based on the data shown in Table 1, the height in metres being plotted vertically and the age in years horizontally. The crosses on the graph correspond to the data

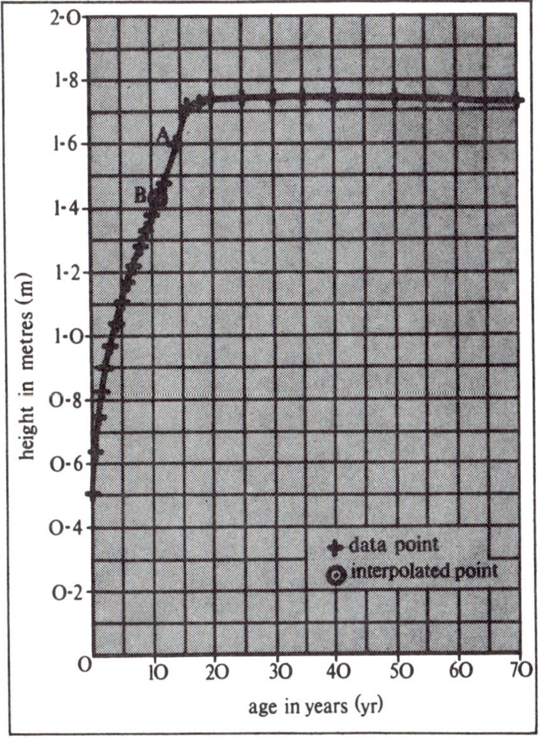

Fig. 1.1. Relationship between height in metres (m) and age in years (yr) for a typical human being from birth to age 70 yr (linear scales).

shown in the table (e.g. cross A corresponds to the measured height of 1·60 metres at age 14 years), while the smooth line drawn through the crosses enables us to visualize the overall way in which one quantity (height) varies with respect to the other (age), and to infer the corresponding values of the quantities between the measured data values shown in the table. For example we can infer from point B that the height is probably about 1·43 metres at 11 years although we have no measurement at this age. The scales along the edges of the graph are usually called the horizontal and vertical axes of the graph and should always be clearly labelled so as to show the quantity represented and the units in which it is measured. The units in this case are metres, abbreviated to m, for the heights, and years, abbreviated to yr, for the ages. In general in this book abbreviated forms for units will be used, although whenever a new unit is

Age yr	Height m	Age yr	Height m
0	0·51	12	1·48
0·5	0·64	14	1·60
1	0·75	16	1·71
1·5	0·83	18	1·73
2	0·90	20	1·74
3	0·97	25	1·74
4	1·04	30	1·74
5	1·11	35	1·74
6	1·17	40	1·74
7	1·22	50	1·74
8	1·28	60	1·73
9	1·33	70	1·73
10	1·38		

Table 1. Relationship between height in metres (m) and age in years (yr) for a typical human being. (Data for Fig. 1.1, 1.2 and 1.3.)

introduced it will first be written out in full and its recognized abbreviation shown in brackets. Appendix 2 contains a list of the units used together with their recognized abbreviations. The whole diagram is said to show the values of one quantity (height) as a function of another (age), and similar diagrams can be used to show the relationship between any two interdependent quantities. The line in Fig. 1.1 shows three distinct types of behaviour. From birth to about 15 yr it is approximately a straight line and the height is said to vary almost linearly with age. From about 15 yr to 20 yr the line curves sharply and the height is said to vary non-linearly with age, while from 20 yr to 70 yr the line is almost a horizontal straight line which means that the height is approximately constant or independent of age.

Fig. 1.2 shows part of the same data plotted with a different, expanded scale along the age axis. The smooth line now looks quite different and it is easier to see that the height does not vary with age in an exactly linear fashion in the age range 0-14 yr. Also the variation of height in the first couple of years is shown more clearly than in Fig. 1.1.

The scales used along the axes in Fig. 1.1 and 1.2 are linear scales

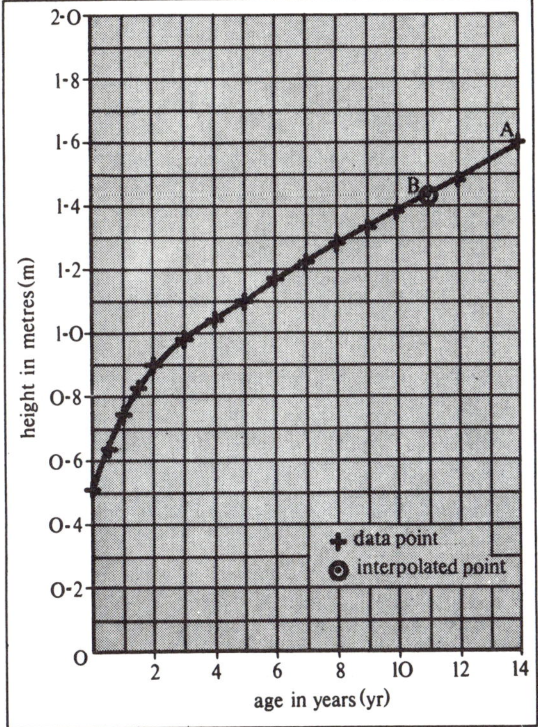

Fig. 1.2. Relationship between height in metres (m) and age in years (yr) for a typical human being from birth to age 14 yr (linear scales).

so that equal distances along the age axis represent equal increases in the number of years; for example, the distance along the age axis in Fig. 1 between the 10 and 20 yr points is the same as that between the 50 and 60 yr points. Fig. 1.3 shows the same data (Table 1) plotted using axes divided so that equal distances along the axes represent equal multiples of the quantities represented; for example, the distance along the age axis in Fig. 1.3 between the 1 and 2 yr points is the same as that between the 2 and 4 yr points, the 4 and 8 yr points and so on. Such non-linear scales are called *logarithmic scales* and are very often more convenient and useful than linear scales when the quantities plotted range from very small to very large values. For example, the age ranges 0·1 to 1 yr, 1 to 10 yr and 10 to 100 yr are given equal emphasis in Fig. 1.3, whereas in Fig. 1.1 the data for the first year or two are almost obscured by compression into too small a region of the graph. An increase of a factor of 10 in a quantity (0·1 to 1 yr, 1 to 10 yr) is called an increase of one order of magnitude, so that the axes of Fig. 1.3 represent two orders of magnitude for height and three orders of magnitude for age. Logarithmic scales can thus be used to show the relationship between two quantities over four, five or many more orders of magnitude in a much more concise manner than if linear scales are used.

Graphs are a very useful way of representing large amounts of data concisely, but they must be used with care. At first glance the lines in Fig. 1.1, 1.2 and 1.3 look very different although they in fact represent exactly the same height-age relationship. Therefore any graphical representation of a relationship must be studied carefully in order to see exactly what is represented along the axes, what units are being used, the range of values shown and the kind of scale employed.

As we have seen, the smooth lines drawn through the data points enable us to infer approximately how the quantities represented vary between the measured values; that is, we can *interpolate* between data points with some confidence. It is much less easy to infer how quantities will vary beyond rather than between measured values—that is to *extrapolate* rather than to interpolate. For example if we were given Fig. 1.2 alone we might be tempted to infer that the height of a human being would continue to increase with age in approximately the same way as it does in the first 14 years so that the height would reach around 2·50m at age 30 and 4·50 m at age 60. However,

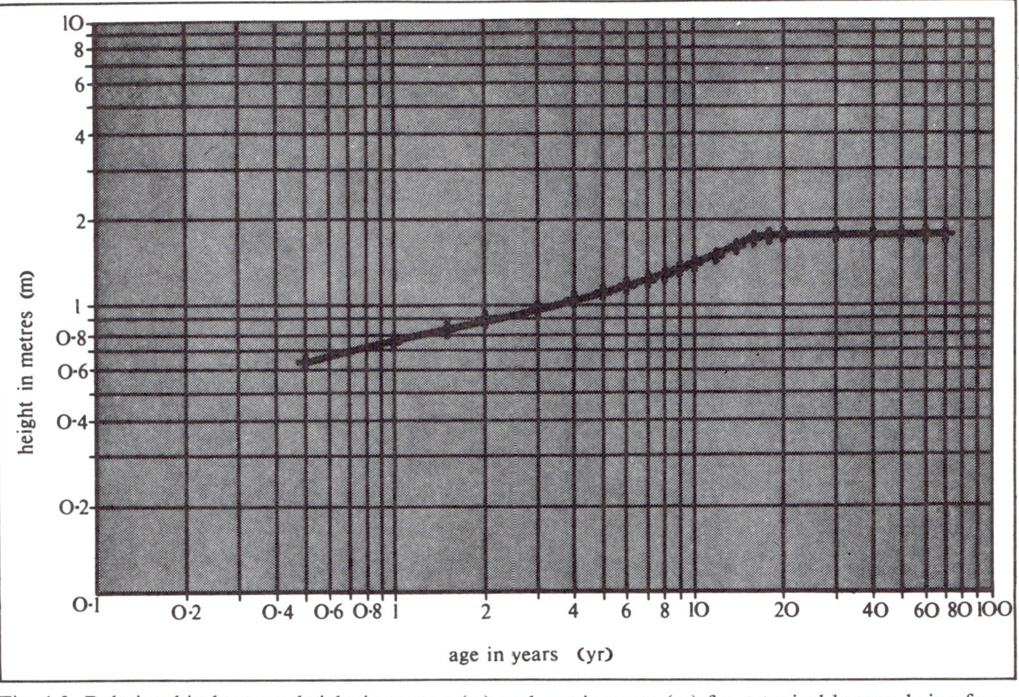

Fig. 1.3. Relationship between height in metres (m) and age in years (yr) for a typical human being from birth to age 70 yr (logarithmic scales).

Fig. 1.1 and 1.3 show clearly that this is not so; the trend of the first 14 years only continues for a few more years and then a new trend takes over. This example shows the dangers of extrapolating even a short way beyond measured values. It is clear that extrapolation should be used with care and we should not be unduly surprised if totally different trends arise, particularly if we extrapolate by many orders of magnitude.

Graphs can also be used in another way which is sometimes useful—namely to represent definite mathematical relationships. Consider Table 2 which has zero and the first four whole numbers in its first column while its second column shows these numbers multiplied by themselves. The numbers in column 2 are called the *squares* of those in column 1, while those in column 1 are called the *square roots* of those in column 2 (9 is the square of 3, while 3 is the square root of 9). This table could easily be extended by more calculations to any whole number or indeed to any number at all. For example, $1{\cdot}5 \times 1{\cdot}5$ is 2·25, so that the square of 1·5 is 2·25 and the square root of 2·25 is 1·5 (the square of a number, say 1·5, is usually written $1{\cdot}5^2$, so that we can write $1{\cdot}5^2$ is equal to 2·25 or, verbally, one point five squared is equal to two point two five). However, it is much easier and more concise to represent the information in such an extended table by using a suitable graph. Fig. 1.4 shows the data from Table 2 plotted with linear scales along the axes; the smooth line drawn

Number *or square root of Number*	Square of number *or Number*
0	0 (0×0)
1	1 (1×1)
2	4 (2×2)
3	9 (3×3)
4	16 (4×4)

Table 2. Relationship between a number (column 1) and the square of that number (column 2) *or Relationship between a number* (*column 2*) *and the square root of that number* (*column 1*). (Data for Fig. 1.4 and 1.5.)

through the data points shows the relationship for intermediate values. Fig. 1.4 can be used either to obtain the square of any number by finding the number itself along the horizontal axis and then reading off the corresponding number along the vertical axis (6·25 is the square of 2·5) or to obtain the square root of any number by finding that number along the vertical axis and then reading off the corresponding number along the horizontal axis (2·5 is the square root of 6·25).

This graph for obtaining squares and square roots can be put in a more useful form by using logarithmic scales (Fig. 1.5). Now the relationship is represented by a straight line and the axes and the

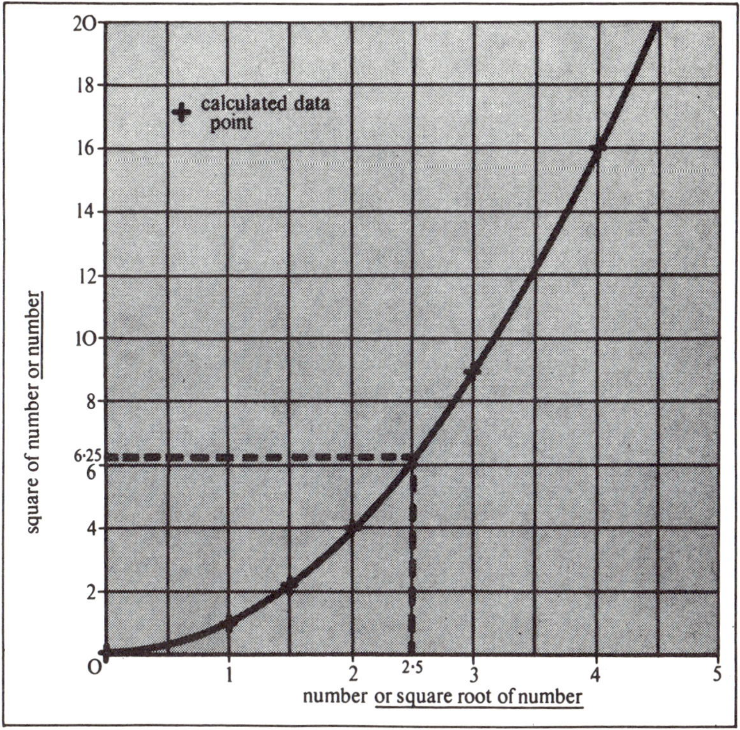

Fig. 1.4. Relationship between a number and the square of that number *or Relationship between a number and the square root of that number* (linear scales).

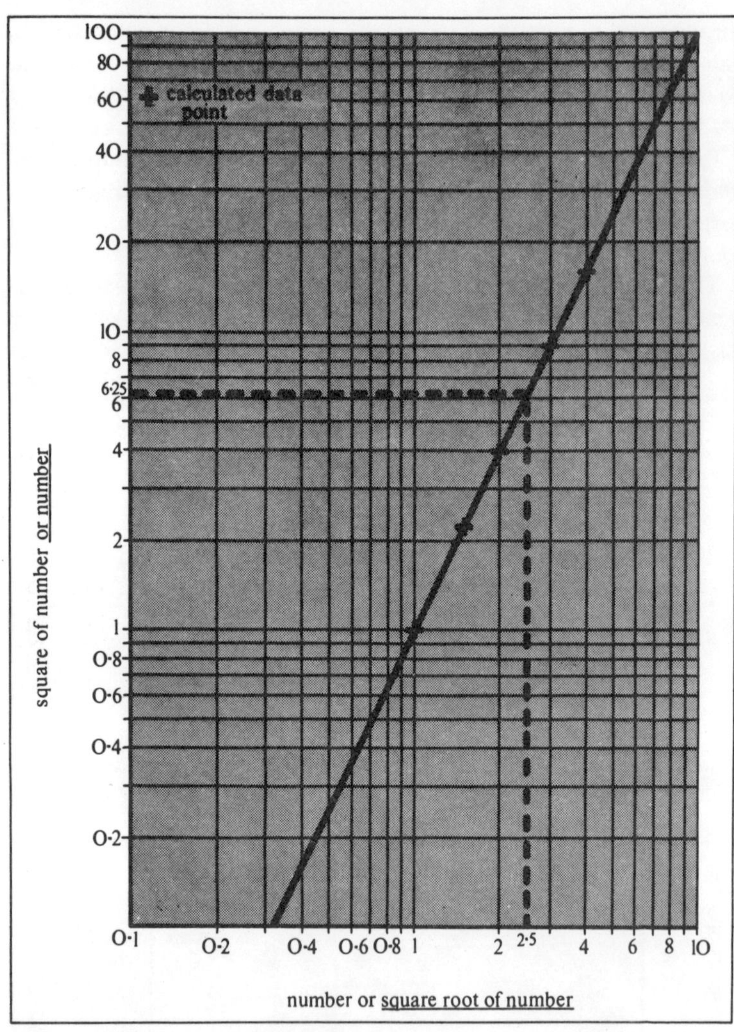

Fig. 1.5. Relationship between a number and the square of that number *or Relationship between a number and the square root of that number* (logarithmic scales).

straight line itself can be extended by many orders of magnitude if required. In this case we can extrapolate with confidence since the relationship is defined precisely for any number and is represented by a straight line. The horizontal and vertical axcs in Fig. 1.4 and 1.5 can obviously be labelled respectively either 'Number' and 'Square of number' or 'Square root of number' and 'Number'. In a similar way, other mathematical relationships can also be represented graphically.

Graphical methods can also be used to specify the positions of various objects, as for example in the use of a grid on a map. Fig. 1.6 shows such a grid which could be drawn on a map of a small region of the earth's surface (small enough for the curvature of the

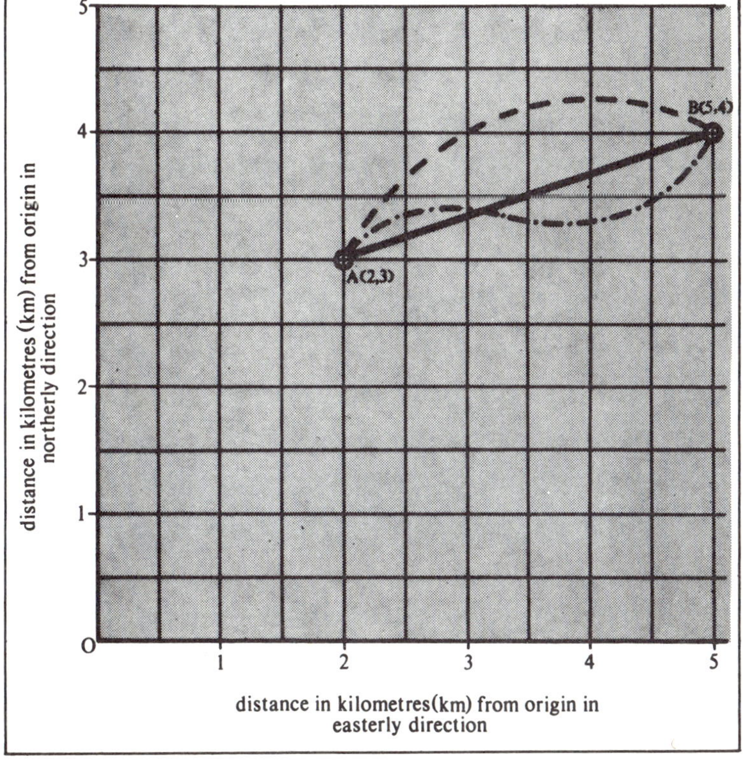

Fig. 1.6. Map grid used to define positions on a map (the lines joining A and B will be discussed in Chapter 7).

earth's surface to be negligible). Consider a town A which can be reached by moving 2 kilometres (km) east from the starting point or origin of the two axes and then moving 3 km north. The position of A with respect to the origin and the easterly and northerly axes of the grid could be represented by the pair of numbers (2, 3). This pair of numbers are called the *coordinates* of A with respect to the origin and the specified axes. Similarly a town B would have coordinates (5, 4) with respect to the same axes and origin, while the origin would obviously have coordinates (0, 0). In this way the position of anything shown on the map can be represented by a pair of numbers or coordinates.

Graphical methods can also be used to specify the successive positions taken up by a moving object. In this case we will restrict ourselves to considering the motion of an object which moves in a particular direction along a straight road since the resulting graphs are then fairly easy to draw and interpret. Consider an object which is at the origin at a certain time and which moves due north at a constant speed of 1 kilometre per hour (km/h). If we take the instant when the object is at the origin as the zero from which times are measured, then the position of the object at subsequent times can be represented by the straight line I in Fig. 1.7 in which elapsed time in hours (h) is plotted along the horizontal axis and distance in kilometres (km) moved in a northerly direction is plotted along the vertical axis. The object has travelled 1 km after 1 h, 2 km after 2 h and so on. Such a graph is called a position-time graph. If the object moves at 2 km/h it will be represented by the straight line II (2 km in 1 h, 4 km in 2 h etc.) and if it moves at 0·5 km/h it will be represented by line III (1 km in 2 h, 2 km in 4 h etc.). An object which is stationary at a point say 3 km north of the origin will be represented by the horizontal straight line IV. Thus any object which moves at a constant speed (including zero speed) will be represented in the position-time graph by a straight line. Furthermore the greater the speed of the object then the greater will be the slope or inclination to the horizontal of this straight line. Line IV is horizontal or has zero slope while lines III, I and II have progressively greater slopes and represent progressively greater speeds. The *slope* of a straight line drawn on any graph is quantitatively defined as the difference between the vertical coordinates of any two points on the line divided by the difference between the horizontal coordinates of the same two points. For example consider line II which passes through the points (1, 2) and (3, 6). The difference between the vertical coordinates of these two points is (6—2) or 4 km while the difference between the horizontal coordinates is (3—1) or 2 h so that the *slope* of line II is 4 km divided by 2 h which is equal to 2 km/h. The same value for the *slope* is obtained no matter which two points on the line are chosen. The *slopes* of lines I, III and IV can be found in a similar way and are respectively 1 km/h, 0·5 km/h and zero. Thus in this case of a position-time graph, the slopes of the straight lines are exactly equal to the speeds of the various objects. This quantitative definition of the slope of a straight line will be used fairly frequently in later chapters.

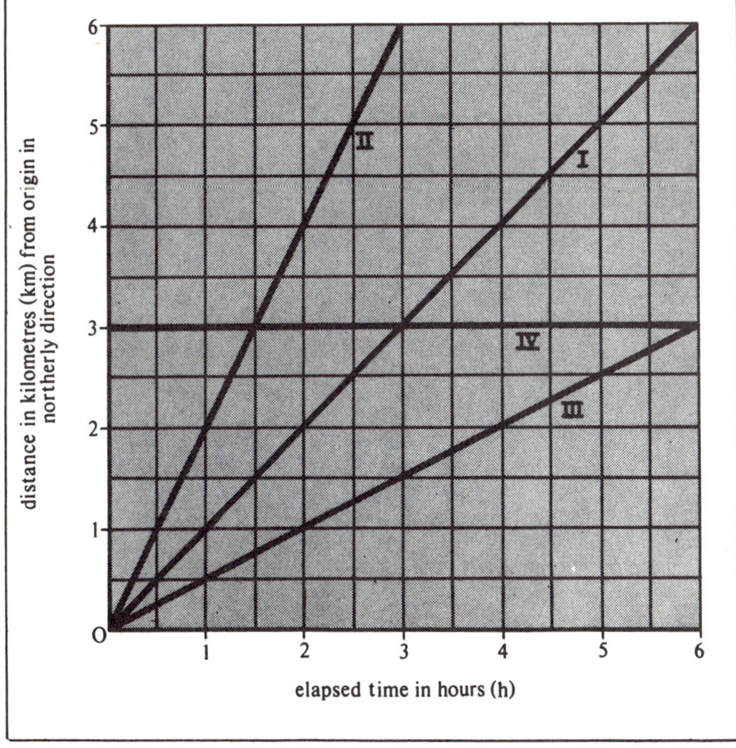

Fig. 1.7. Position-time graphs. Lines I, II, III and IV represent respectively objects moving with constant speeds of 1 km/h, 2 km/h, 0·5 km/h and zero.

In the last paragraph we have considered the position-time graphs of objects moving at constant speeds. If the speed of an object is changing, the position-time graph will not be a straight line but will be curved in some way. For example, if the object starts from the origin at rest and then moves at a varying speed, its position-time graph might resemble line I in Fig. 1.8. The slope of such a curve

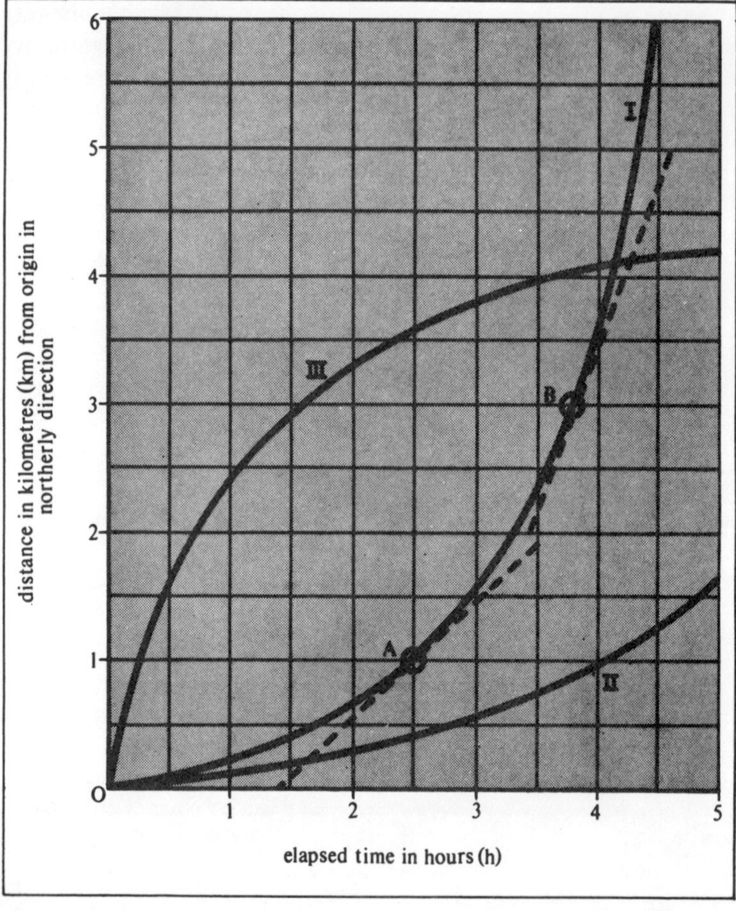

Fig. 1.8. Position-time graphs. Lines I and II represent objects moving with positive accelerations; line III represents an object moving with a negative acceleration.

at any point can be found by drawing a tangent to the curve at this point and measuring its slope. Thus the slope of line I is zero at O, small at A and larger at B corresponding to zero speed at O, a small speed at A and a larger speed at B. An object moving in such a way is said to be accelerating or to undergo an acceleration. This acceleration is greater if the speed of the object is changing rapidly than if it is changing slowly, so that line II in Fig. 1.8 represents an object moving with a smaller acceleration than line I. The acceleration is obviously greater if the position-time graph is more tightly curved or, in other words, has a greater curvature. Line III in Fig. 1.8 is curved in the opposite sense to lines I and II and represents an object whose speed is decreasing with time. Such an object is said to be decelerating or to undergo a negative acceleration. Position-time graphs will be used frequently in later chapters to explain some of the ideas involved in both classical and special relativity.

We saw in Fig. 1.1, 1.2 and 1.3 that a graph can be used to show the way in which one quantity (human height) varies with respect to another (age or time). A useful extension of this method of representation can be made when there are two separate quantities which are varying with respect to a third quantity and we are interested in how the product (one quantity multiplied by the other) of the quantities varies.

Suppose we have graphs which show how the total population and the average annual personal income per head of the population of a country vary over a period of time. Fig. 1.9 (a) and 1.9 (b) show such graphs for a hypothetical country where the population and the average income increase linearly with time over the period considered; it is of course extremely unlikely that such simple linear relationships would actually exist in any real country. The total of the annual personal incomes for the whole country at any time is obviously given by the product of the population and the average annual personal income per head. Hence we can see how this total varies with time by multiplying the population in any given year (from graph (a)) by the average personal income in that year (from graph (b)) and plotting the result on a third graph (Fig. 1.9 (c)). Corresponding points in graphs (a), (b) and (c) are labelled with the same letters. It can be seen from graph (c) that this total does not vary linearly with time.

Such a point-by-point multiplication of two graphs to give a third

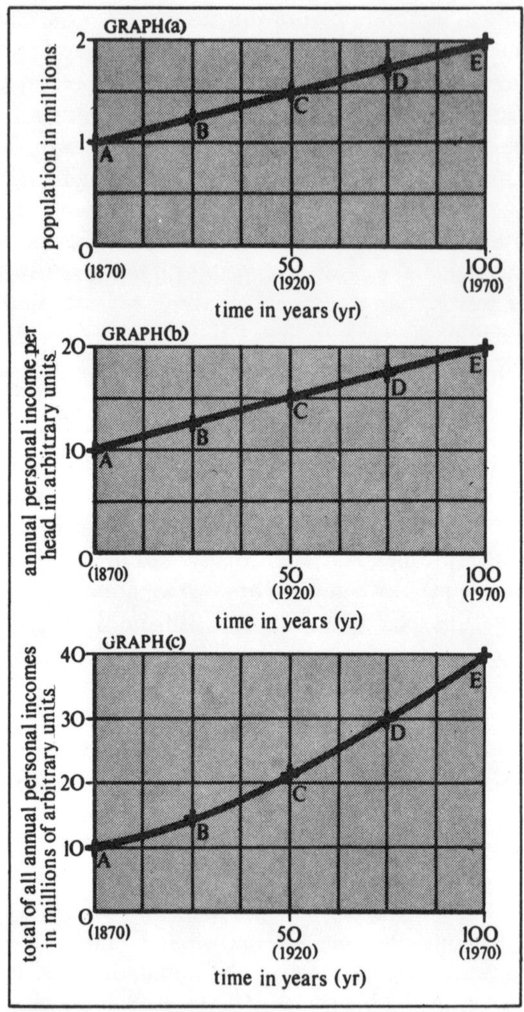

Fig. 1.9. Graphs to illustrate the point-by-point multiplication of two graphs (Graphs (a) and (b)) to give a third graph (Graph (c)).

graph which shows how a product varies will be useful in later chapters; it can of course only be done if one of the axes (here the horizontal or time axis) is the same in all three graphs. We have now developed most of the main mathematical or quantitative ideas which we shall need and we can now turn to the third common difficulty mentioned at the beginning of this chapter.

This third difficulty about the ideas of relativity, that they appear to fly in the face of common experience or common sense, can be shown to be only an apparent difficulty if we consider how restricted our normal environment is. Commonsense ideas are developed during the early formative years of life which are usually spent in one particular region of the earth's surface. Also the commonsense environments of different people or peoples differ widely in climate, culture and language, so that what is normal or commonsense behaviour in say London in 1970 (speaking English, spending pounds or carrying an umbrella) would be entirely inappropriate or abnormal behaviour in an Eskimo settlement north of the Arctic circle or in an agricultural commune in the Chinese countryside or even in what was London in the year AD 70 or 1070.

These differences are already fairly wide, but all such earthly environments are restricted in many more ways than this. They are restricted in space to distances or sizes of material objects ranging from about a millimetre to a few thousand kilometres, in time from about a tenth of a second to a few years and in temperature from about —40°C to 50°C. However, the environments in which the scientist works, either when using special laboratory apparatus or when studying the earth and the universe, extend far beyond these limits. His distance scale extends down to distances as small as one thousand million millionths of a metre (0·000000000000001 or 10^{-15} m) in nuclear physics and up to distances as large as 10 million million million million metres (10,000,000,000,000,000,000,000,000 or 10^{25} m) or more in astronomical observations of distant stars or galaxies of stars. His time scale extends down to times as short as one thousand millionth of a second or less (0·000000001 or 10^{-9} s) in atomic and nuclear physics and up to about five thousand million years (5×10^9 yr or $1{\cdot}5 \times 10^{17}$ s) in geological work and even further in theories concerning the origin of the stars and galaxies. His temperature scale goes down to about —273°C and up to 20 million degrees or more in the interiors of stars. Thus the distances, times and temperatures with which the scientist works may exceed or fall below those with which common sense is familiar by many orders of magnitude.

We saw earlier that extrapolating for even one order of magnitude beyond the limits of experience (Fig. 1.2) could be misleading; how much more misleading might it be if we extrapolate for five, ten or twenty orders of magnitude. It is therefore not surprising that new phenomena, unknown to common sense, are observed by scientists. The really surprising thing is how many commonsense ideas survive when we step so far beyond the limits of the commonsense environment.

Our normal environment is limited in two other ways which have a bearing on relativity. First, we are usually limited to speeds which range from zero to about 1 km/s (the speed of a typical supersonic aircraft) whereas relativity deals with speeds up to about 300,000 km/s. Secondly, in our earthly environments we are always subject to the gravitational pull or attraction of the earth which gives rise to such fundamental commonsense ideas as 'up' and 'down'. Space travel has now introduced most people to the idea of environments in which there is virtually no gravitational effect and therefore no 'up' or 'down' in the usual sense of these words. We shall be concerned later in the book with how things behave in different gravitational circumstances.

It is thus not difficult to see intellectually how limited our commonsense environment is but it is much more difficult to accept these limitations intuitively and to incorporate them completely into our ideas of the world. This is because even a practising scientist still spends most of his everyday life in the commonsense world and his commonsense experience far outweighs what might be called his extra common-sensory experience. If, therefore, in later chapters we encounter ideas or phenomena which our common sense cannot or does not accept, it will be useful if we do not reject the new ideas or phenomena but ask instead, 'Is our common sense qualified to judge in this case?'

2

The Nature of Theories and Experiments

Men's ideas about the nature of the physical world have nearly always developed by means of both theories or general speculations and experiments or observations, even if the observations have only been the implicit ones involved in appeals to common sense. For a long time these two approaches to understanding nature tended to be largely separate. For example there were the mainly verbal theories of the Greeks about the basic constituents of matter in which experiment played little part, and the mainly experimental work of the early alchemists or of engineers of all kinds in many different civilizations all over the world. Then at some time early in the sixteenth century there began in Europe a development which led to the extremely rapid growth of science which has continued to the present day and which has had such an all-pervading influence on the modern world.

This development involved two main tendencies which are important for our purposes. One was the tendency for theories of the physical world to be expressed in ways which were more precise, quantitative or mathematical than had been common in the mainly descriptive or qualitative theories of earlier times. The other was the tendency for theory and experiment to be much more closely linked and therefore to influence one another much more than they had in the past. Why these two tendencies, which had such momentous consequences, should have developed in Europe just at this time and not for example in China, which was in many ways a technologically more advanced civilization, is a fascinating question which historians are still discussing. However, it is outside the scope of this book and it is enough for us to look a little more closely at the interaction between theory and experiment in the development of our ideas of the physical world.

A simplified picture of this interaction has often been given which involves an alternation of experiment and theory. In this picture the first source of scientific knowledge about a particular topic is an experiment or series of experiments carried out under carefully controlled and documented conditions. It is usually implicitly assumed here and in all scientific work that the results of similar experiments carried out under otherwise identical conditions but at different places or times are identical (in other words, the experiments are repeatable). After the initial experiments have been carried out, scientists attempt to construct a theory which incorporates all the known experimental data amongst its predictions and also predicts some new unobserved phenomena. Further experiments are then carried out to search for these phenomena and the results of these experiments lead to the confirmation, rejection or, more likely, modification of the theory. The modified theory then suggests further experiments and so the process continues.

This outline of how science develops contains some grains of truth but in fact the interaction between theory and experiment is more complicated, close and continuous than this in most fruitful scientific work. It will be easier to understand this close interaction and to appreciate some of the significant points involved in the theories of relativity if we now look in more detail at the nature of scientific theories and experiments.

A scientific theory is an attempt to combine together a large number of experimental observations into a systematic scheme of general ideas or relationships. One important reason for trying to find such general relationships is that the number of experimental observations is so large that it is difficult or impossible to comprehend them all without the help of some unifying theoretical ideas.

Theories in the physical sciences are almost always mathematical and usually involve four basic processes. First the quantities involved in the theory are defined both qualitatively (in words) and quantitatively (by some form of mathematical equation); such quantities are sometimes simple concepts such as the position or speed of an object, which we have already mentioned in Chapter 1, and sometimes more complex concepts such as *force* and *energy* which we shall deal with in Chapter 3. The next step is to obtain or formulate mathematical relationships between these quantities. This step usually involves making some simplifying assumptions about the experiments which the theory is seeking to describe.

The third step is mathematical and consists of rearranging the mathematical relationships into a different but equivalent form. This step often makes clear some important consequences of the relationships which were implicit but not obvious in the original formulation. The final step is to give a physical or experimental interpretation of these new mathematical relationships. This step may also involve some simplifying or restrictive assumptions which then form an integral part of the theory. The third step, the mathematical manipulation, is sometimes considered to be the most important or significant part of a theory. Indeed it is sometimes presented as if it alone represented the theory, so that the other steps are either ignored or regarded as of limited interest and importance. In fact the other steps are of equal or greater importance and are sometimes considerably more difficult to carry out.

Very often it is only when some simplifying assumptions have been made and a considerably simplified model of the complicated real world has been constructed that any mathematical analysis becomes possible. It is therefore necessary to have a clear idea of the assumptions or limitations involved in setting up or interpreting mathematical theories so as not to draw misleading conclusions or make false deductions from them.

A good theory should include all known relevant experimental results among its predictions and should also make some new predictions which can then be experimentally tested. If it is a new theory which is trying to replace an older theory, then it is desirable that it should be applicable to a wider range of experiments than the older theory and that it should dispense with some or all of the simplifying assumptions involved in the earlier theory. However, it still has to incorporate the experimental evidence which supported the older theory. Therefore a new theory does not usually demolish an older theory and replace it with something totally new. It rather supersedes the older theory, retaining all its successes while extending its range of applicability or removing some of its inadequacies.

Finally the most crucial point to make about any theory is that it is only as good as its last experimental test. A theory is always on probation and if at any time the experimental evidence is clearly at variance with a theoretical prediction then it is the theory which must be changed. This does not mean that a single point of disagreement between theory and experiment necessarily leads to the total abandonment of a previously successful theory. Rather it means that a new theory should be sought which both incorporates the successes of the old and removes the disagreement. Experiments can therefore invalidate a theory but can never establish a theory as inviolate; a theory to be scientific must always be capable of disproof. The attitude to theories and experiments expressed in this paragraph is that of the distinguished philosopher of science, Karl Popper, and is generally accepted by most working scientists.

Experiment then is the final arbiter between theories as well as their progenitor and thus plays a crucial role in the development of science. Let us therefore now look more closely at the nature of experimental work and some of its limitations.

It is sometimes thought that experimental measurements are always extremely precise and result in exact numerical values for the measured quantities. For example the width of a wooden table as measured with a steel measuring tape might be found to be say 1·211 m. And of course such exact numerical results are highly desirable if experiments are to play the role of arbiter between theories which was mentioned above. Unfortunately many measurements are not quite as precise and definite as this but are much more indistinct and blurred at the edges. Two of the main reasons for this imprecision are that the quantity being measured may itself be changing its value as external conditions change and that any measurement is subject to errors of various kinds. In our example the wooden table may swell and change its width as the humidity increases, while the measuring tape may change in length as the temperature rises or may have been calibrated (compared with a standard of length) with limited accuracy.

Changes in the quantity being measured can usually be avoided by specifying the conditions of measurement more closely (in our example by specifying the exact humidity which existed when the measurement was made) but the difficulties caused by the measurement apparatus will still give rise to what are called systematic errors in the measurement. Thus either a rise in temperature or an error in calibration which led to the marks on the tape being too far apart would lead to consistently or systematically low values for measured lengths. In this simple case it is not difficult to allow for the first systematic error if we measure the temperature and know how the tape material expands as the temperature rises, while an estimate of the other systematic error can usually be obtained from the tape manufacturer. In many measurements, however, it is not very easy to locate all sources of systematic error and even less easy to estimate their size.

A further type of error that is even more important in some measurements can arise when a measurement is being made in the presence of other phenomena which are also detected by the measuring instrument. Suppose the strength of a weak radio signal from a distant station is being measured. The output from the loudspeaker probably includes a certain amount of interference or noise as well as the signal of interest so that any measurement of the total output is in error due to this interference. In many cases such interference or noise is just as likely to increase as to decrease the strength of the measured signal and so is said to give rise to a random error in the measurement. Such random interfering signals can sometimes be reduced by careful design of measuring apparatus but in most experiments they provide a definite limit to the accuracy with which measurements can be made.

Thus any numerical measurement should always be accompanied by an estimate of its accuracy so that the weight to be attached to the measurement as evidence for or against a theory can be assessed. If the accuracy is high (very small error) then the measurement would have great weight, whereas if the accuracy is low (large error) then the measurement would not be given such great weight. Another consequence of this limited accuracy of experimental measurements is that theories are always potentially vulnerable to advances in experimental technique which increase the accuracy with which measurements can be made. A theory which today has only a tentative status because its crucial predictions cannot be tested to a sufficient degree of accuracy may tomorrow, because of such an advance, be either confirmed or discredited.

It should now be clear that both theories and experiments are not quite so clear-cut and precise as they were assumed to be in the simple picture of the development of scientific ideas which was described at the beginning of this chapter. Also we can now introduce some general points about scientific theories which are relevant to the discussion of Einstein's theories in later chapters.

First let us consider the disputes which exist between competing theories in most scientific fields. These disputes have often been puzzling to the non-scientist who has perhaps assumed or been brain-washed to think that the adjectives scientific and objective are virtually identical. It should now be clear how such disputes can arise quite legitimately. Experiments which arbitrate unambiguously between theories need to be performed with great accuracy and this is often hard to achieve. Another difficulty is that corrections often have to be made to experimental observations before they can be properly compared with theoretical predictions. Such corrections may be due to some known inadequacy in the experimental technique, as in the example of the measuring tape expanding with increasing temperature, or may be an attempt to allow for the only partial correctness of some of the assumptions made in developing the theory. Also there may be corrections which really ought to be applied but which for some reason have not been made. And finally the values of the corrections themselves will not be known precisely and this imprecision will therefore introduce some additional error into the measurements.

For all these reasons the experimental evidence for or against a particular theory may be conflicting, and different scientists may evaluate the same experimental observations in different ways. Such disputes are fairly common because it is precisely the disputed or uncertain areas of a subject which are most interesting and most likely to give rise to fruitful new results.

Another interesting possibility occurs if the differences between the predictions of two theories are appreciably less than the errors of the measurements or corrections involved. Then the proper conclusion to draw is that the theories are indistinguishable or equivalent while the present measurement techniques remain unimproved. More subtle cases of equivalence between different theories arise when the differences between the predictions of two theories are negligible (much less than the measurement errors) under some conditions but not under others. Fig. 2.1 shows the way in which the

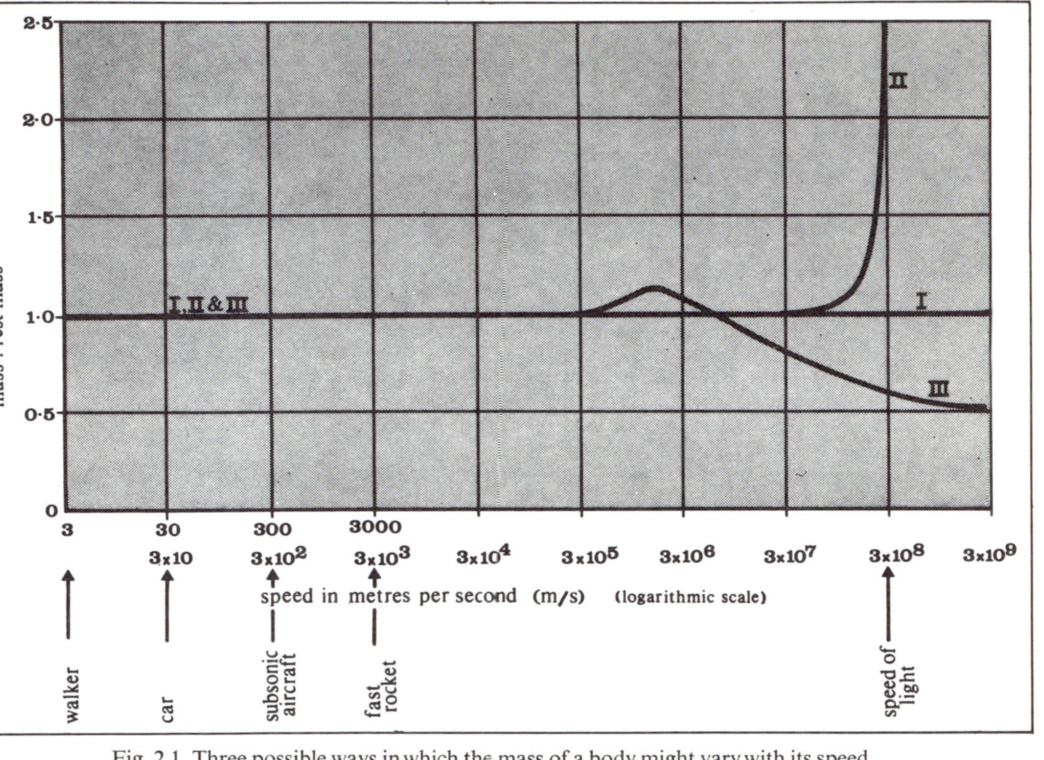

Fig. 2.1. Three possible ways in which the mass of a body might vary with its speed.

mass of a body (see Chapter 3 for a precise definition of *mass*) varies as its speed changes according to two different theoretical viewpoints. The mass of the body divided by its mass when it is at rest (or *rest mass*) is plotted on a linear scale along the vertical axis while the speed in metres per second (m/s) is plotted on a logarithmic scale along the horizontal axis. Curve I shows the variation or rather the lack of variation which would be expected if the ideas about mass deduced from terrestial and astronomical measurements up to the end of the nineteenth century (pre-Einstein) were valid for all speeds. Curve I is a horizontal straight line which implies that the mass is constant at all speeds. Curve II shows the variation of mass with speed which is predicted by Einstein's special theory of relativity; curve II almost coincides with curve I at low speeds but the two curves clearly diverge as the speed increases.

In the range of speeds which constitute our commonsense experience (up to about 3×10^3 m/s, the speed of a fast space rocket) the maximum divergence between the two curves is about one part in a million. Thus in this range of speeds the theories are equivalent in the sense discussed above since it is not experimentally feasible to detect such small mass changes in bodies moving at such speeds. For speeds of about 3×10^7 m/s the two curves differ by about one part in a hundred (1 per cent or 1%) while as the speed approaches 3×10^8 m/s the curves diverge rapidly and curve II shows the mass increasing without limit (or tending to an infinite value).

The experimental evidence is strongly in favour of curve II, as will be discussed in Chapter 5, but the important point here is that it is possible for two theories to be indistinguishable or equivalent in a certain range of conditions (in this case in the range accessible to common sense) and yet to differ widely outside that range. Two curves such as curves I and II are said to converge or become *asymptotic* to each other at low speeds and it is possible to say that the theories underlying the curves are also *asymptotic* to one another for these speeds. Curve III shows another possible mass variation which is also asymptotic to curves I and II at low speeds but which diverges from these curves at higher speeds. However, as far as I know, curve III does not represent any theory which has actually been proposed.

It should now be clear from the general discussion of scientific theories in this chapter that any new more general theory which seeks to supersede an accepted theory must be asymptotic to the accepted theory over the range of conditions for which the accepted theory has been found satisfactory. This conclusion will be very useful in later chapters when we are considering the relationship between Einstein's relativity theories and the theories which his relativity theories were attempting to supersede.

The final general point about scientific theories which I wish to make arises from that close and almost continuous interaction between theory and experiment which we have already observed to have had such a great influence on the rapid development of science since the sixteenth century. This interaction has been developed a stage further in the twentieth century with the introduction of what is sometimes called the operational viewpoint into the construction of theories. This viewpoint requires that any concept or quantity which is used in a scientific theory should be defined operationally—that is, it should be defined in terms of definite physical operations or measurements. This operational requirement should be applied to any concept or quantity no matter how obvious or apparent to common sense the nature of that concept or quantity may be.

As a very simple example, suppose that we want to develop a theory which enables us to predict the motions of the sun, the earth and the moon over a period of time. As a starting-point we would probably need to know the relative positions of the three bodies at some initial time on our terrestrial clocks. The operational viewpoint would require us to specify exactly how we would carry out such measurements, making any necessary corrections such as allowing for the finite time light takes to reach us from these bodies (about 8 minutes from the sun and 1·25 seconds from the moon). It would not be sufficient just to assume that such measurements could be made.

The operational viewpoint is not usually put forward as an absolute requirement but as a desirable one. Any theory set up on this basis would have a greater chance of being soundly based and would be less likely to need subsequent serious revision than a theory which involves concepts or quantities which cannot be operationally defined. The importance of the operational viewpoint was clearly appreciated by Einstein, and it was one of the fundamental ideas behind his special relativity theory (see Chapter 4). Also, as we have seen, it is a natural extension of a trend in science which had been apparent for three hundred years or so.

The discussion of the nature of theories and experiments in this chapter has taken us a little further forward compared with our position at the end of Chapter 1. We should now be able to see that when we move out from our commonsense environment we should expect that things might be different but different in a special way. They might be different but, if they are, they must also be asymptotic to our commonsense ideas. So far, however, we have no way of knowing the form such asymptotic ideas might take. In the next chapter we shall discuss in more detail the accepted ideas about the physical world which are valid in the realm of commonsense experience, while in the rest of the book we shall outline just what different but asymptotic ideas were introduced by Einstein's theories.

3

Classical Ideas in Physics: Relativity Before Einstein

Classical physics is the term usually used to describe the body of knowledge, theoretical and experimental, about the physical world which had been established by the end of the nineteenth century. At this time our knowledge about such topics as matter and motion, heat, light, sound, electricity and magnetism appeared to be reaching a reasonably satisfactory state in the sense that most experimental phenomena could be fairly well explained in terms of general theories. Some scientists were even heard to regret that all the major discoveries had been made and that the only tasks remaining for them were fairly trivial. They were hopelessly mistaken for in only a few years the introduction of totally new concepts such as relativity, quantum theory and wave mechanics was to transform physics almost beyond recognition. And yet the establishment of the ideas of classical physics was in itself a tremendous achievement, which involved a number of extremely difficult concepts. Many of these concepts are now presented to people at a time in their lives when their receptivity to new ideas is high and their critical powers are not yet fully developed. As a result the importance and difficulty of these concepts is not always fully appreciated.

In this chapter the ideas of classical physics which are needed later in the book will be introduced. This will be done partly in order to indicate and stress some of the difficult classical concepts which are retained in Einstein's theories and partly to establish clearly the previously accepted theory to which any new and more general theory must be asymptotic. In passing let us note the time scale of the physicist who uses the term classical physics to refer to the state of knowledge of the physical world which had been achieved some seventy years ago, whereas when the artist or historian refers to the classical period he is looking back two thousand years or so to the civilizations of Greece and Rome. Perhaps the physicists are using the term classical in its other sense—that which is of the highest rank or importance and hence deserving of emulation.

The fundamental ideas of classical physics are those of space and time, and we have already mentioned in Chapter 2 that in most of science it is implicitly assumed that the results of experiments will be the same no matter where or when they are performed, provided of course that all other conditions are identical. In other words experiments are assumed to be repeatable and their results to be independent of the position in space or time at which they are performed. To use the technical jargon, experimental results are assumed to be *invariant* with respect to different positions in space and time. This assumption, which is retained in Einstein's work, is involved whenever we check an experimental result by repeating the experiment, whenever we make use of the experimental results of other scientists working at different places and at different times, and whenever we attempt to apply established physical laws, determined mainly by terrestrial experiments, to the universe as a whole. It is an assumption without which it would be extremely difficult for science to progress and its importance underlines the importance of the ways in which positions in space and time are specified. The idea of invariance is also important and will recur both in this and later chapters.

Positions in space are commonly described by a simple extension of the sort of map grid which was discussed in Chapter 1. There the position of a point on a flat surface was described with respect to an origin and two perpendicular axes by two numbers or coordinates. Such a flat surface is said to be two-dimensional since two axes and two coordinates are required to describe positions on the surface. The position of a point A in the three-dimensional space above a flat horizontal surface can similarly be described if we imagine a straight line drawn from the point A to a point B on the flat surface vertically below A. The position of the point A can then be described by three numbers, the two coordinates of B together with a third number or coordinate, the height of A above B, which of course is the length of the line AB. These three coordinates in effect specify the position of A in terms of the distances of A from an origin O along each of three mutually perpendicular axes (see Fig. 3.8 (a)). Such systems of three mutually perpendicular axes are usually called cartesian axes and the resulting sets of three coordinates called cartesian coordinates after the mathematician and philosopher Descartes who first used them extensively. Cartesian coordinates are not the only type of coordinates which can be used to specify positions in space but they are by far the most commonly used.

In addition to a coordinate system, the specification of positions in space requires the establishment of a standard of length so that distances along the axes can be measured. The accepted international standard for many years was the length of a metal bar which was kept in Paris under suitable controlled conditions. This was called the standard metre and all measurements of length were indirectly referred to it. Advances in measurement techniques have recently led to the adoption of a new length standard, the wavelength of the light emitted by a particular kind of atom, which has been found to be a more stable and reproducible standard than the metal bar (see Appendix 2).

The distance between two points in space in terms of the coordinates of the points is another quantity which is sometimes required, and this can be calculated fairly easily. As a simple example let us revert to a two-dimensional flat surface and let us suppose we want to find the distance between the origin O with coordinates (0,0) and a point A with coordinates (3, 4), the units of length along the axes being metres (see Fig. 3.1). The distance we require is 5 m, as can be found by direct measurement. A general way of calculating such a distance has been known for over two thousand years and is usually attributed to the Greek philosopher Pythagoras. We first draw a triangle OAB as in Fig. 3.1 by joining together the points O, A and the point B with coordinates (3, 0); this triangle obviously has a right angle at the point B while the sides OB and BA, which enclose the right angle, have lengths of 3 m and 4 m respectively. We then calculate the squares of the lengths of the sides OB and BA which enclose the right angle (in this case, the square of 3, or 3^2, is 9, and the square of 4, or 4^2, is 16), add together these squares (9 and 16 make 25), and then find the square root of the result (the square root of 25 is 5). This square root gives the required distance OA as 5 m which is the same value as was obtained by direct measurement.

In this example the numbers involved are all very simple but the same general method can be used for any values of the lengths of the sides with the help of a conversion chart for squares and square roots such as Fig. 1.4 or Fig. 1.5. If the lengths were say 1·5 m and 2·5 m then Fig. 1.4 gives the square of 1·5 as 2·25 and the square of 2·5 as 6·25. The sum of the squares is thus 8·5, and Fig. 1.4 gives the square

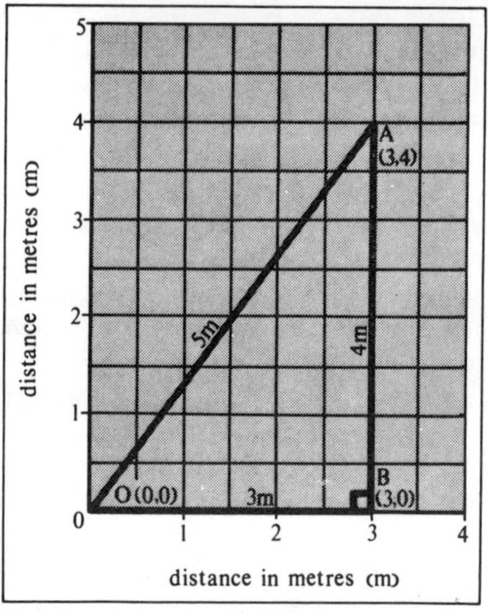

Fig. 3.1. Diagram to illustrate Pythagoras' method for distance calculations. $OB^2+BA^2=OA^2$ ($3^2+4^2=5^2$).

root of 8·5 as 2·9, which is the required distance in metres (the accuracy of this result is limited by the care with which Fig. 4 has been drawn—more accurate methods for calculating square roots exist and could be used).

We have described the application of Pythagoras' method to the problem of finding the distance between the origin and any other point on a flat two-dimensional surface. It is not difficult, however, to see how the same method could be used to calculate the distance between any two points on the surface if we construct a suitable right-angled triangle with the sides enclosing the right angle drawn parallel to the coordinate axes. It is also fairly easy to extend Pythagoras' method so as to be able to calculate the distance between any two points in three-dimensional space if we know the co-ordinates of the points; however we will not discuss this calculation in detail.

Time is measured by making use of phenomena which appear to be periodic or cyclic, such as the daily passing of the sun through its highest point in the sky, the oscillation of the bob of a pendulum or the vibrations of some solid bodies such as tuning forks or quartz crystals. These phenomena are used to construct clocks which measure the time between two particular events by recording in some way the number of cycles or periods which occur between the two events. The difficult question as to whether such periodic phenomena are 'really' periodic and always take the same 'time' to go through one cycle of change is usually evaded by choosing phenomena which give consistent results when the 'times' given by different clocks are compared with one another. The standard of time was for many years the length of the mean solar day. This is the time which elapses between passages of the sun through its highest daily point in the sky averaged over a year, and was defined as being 86,400 seconds. However, difficulties over the use of this standard have recently led to the adoption of a new, more reproducible standard, the period of vibration of a particular type of radiation emitted by a particular atom (see Appendix 2).

Once length and time have been defined, it is easy to define the speed of a moving body as the length traversed in any direction divided by the time taken to traverse this length (see Chapter 1). The *velocity* of a moving body is a more precisely specified quantity; it is defined as including both the speed of the body and the direction in which it is moving. It is therefore incorrect to say that a body has a *velocity* of 3 m/s without specifying its direction of motion. A more correct statement would be that the body has a velocity of 3 m/s along a straight line running say due east from a particular point, or that the *velocity* has a *magnitude* of 3 m/s in that direction. This definition of velocity means that the velocity of a body changes either if its speed changes or if its direction of motion changes or if both change. Any change in the velocity of a body means that the body undergoes an *acceleration*, which is defined as the change in velocity divided by the time during which the change takes place or as the rate of change of the velocity with respect to time. Since the acceleration of a body is related to a change in velocity, it is also a quantity which is only completely defined when a magnitude and a direction are specified; such quantities are called vector quantities or vectors. Velocity and acceleration are widely used in physics and, as we have seen, are defined solely in terms of the basic quantities length and time.

The topics discussed so far in this chapter—the positions, velocities and accelerations of moving bodies and the relations between them—constitute a subject which is usually called *kinematics.* Kinematics deals quantitatively with motion without asking why bodies have the motions they do or why their motions should change in any way. The subject which both asks and answers these questions about motion is called the *dynamics* of moving bodies and has its roots in the work of Galileo and Newton in the sixteenth and seventeenth centuries.

The first basic idea of *dynamics* as developed by Galileo and Newton is that the natural state of motion of a body is one of constant velocity. That is, a body isolated from its surroundings or left to itself will continue at a constant speed in a given direction indefinitely. (A special case of this natural motion occurs when the constant velocity is zero or the body is at rest.) This means that no cause need be sought to explain a motion with constant velocity (or a state of rest); it is the natural motion of an isolated body. It is only when the velocity of the body changes in magnitude or direction (when the body has a finite acceleration) that a cause need be sought. This statement about the natural motion of a body is sometimes called the principle of *inertia* or Newton's first law of motion, and such natural motion is also sometimes called *inertial* motion.

The second basic idea of *dynamics* concerns *forces.* A *force* is defined qualitatively as being anything which causes a body to move unnaturally or, what is the same thing, to change its velocity and thus experience an acceleration. A *force* is defined quantitatively in terms of both the *mass* of the body and the acceleration of the body produced by the *force.* Therefore it is necessary to define the *mass* of the body before force can be defined, and it too is defined in terms of acceleration in the following way. Suppose we have a force which we know to be constant since it always produces the same acceleration in the motion of a particular body; such a force might be that exerted by a spring which is always extended by the same amount. If this constant force causes a certain body A to experience an acceleration of say 1 metre per second per second (m/s^2) while the same constant force causes another body B to experience an acceleration of say 2 m/s^2 (see Fig. 3.2), then the *mass* of A is defined

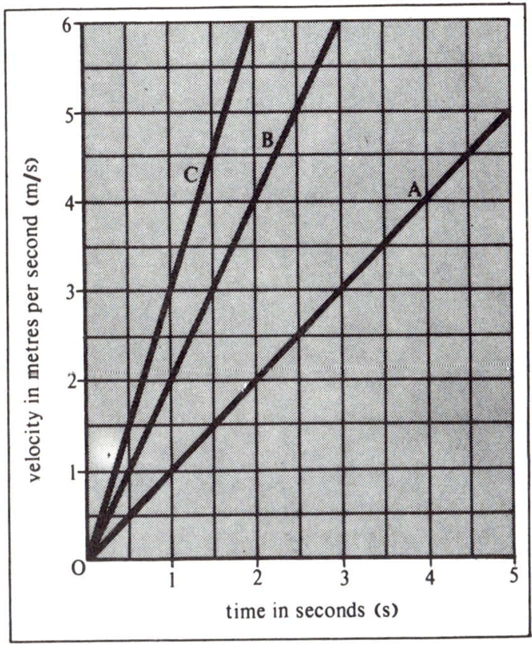

Fig. 3.2. Velocity-time graphs used in the definition of the masses of three bodies A, B and C. All the bodies are initially at rest and are subjected to the same constant force.

as being twice that of B. If a third body C experiences an acceleration of 3 m/s^2 then the mass of A is three times that of C, and so on. In this way the ratio of the masses of any two bodies can be found. The standard of mass is the mass of a certain piece of metal kept near Paris which is defined as the standard kilogramme. So if one of the masses in our experiment is a subsidiary standard which has been accurately compared with the standard kilogramme, then we can find the mass of any other body without knowing the magnitude of the force which produces the accelerations. The mass of any body as we have defined it is sometimes called its *inertial mass*.

Force can now be quantitatively defined by a similar sort of experiment. Suppose a force produces an acceleration of 1 m/s^2 when it acts on a mass of 1 kilogramme (kg); this force is defined as 1 unit of force or 1 newton (N). If another force acts on a mass of 1 kg and produces an acceleration of 2 m/s^2 then it has a magnitude of 2 N, if it produces an acceleration of 3 m/s^2 then it has a magnitude of 3 N and so on (see Fig. 3.3). This definition of force was first made by Newton and it is sometimes called Newton's second law of motion; of course the unit of force is also named after Newton. Force is a vector quantity since it is defined in terms of acceleration which itself is a vector. Also since force is defined in terms of acceleration, which itself is defined directly in terms of length and time, force too depends on the fundamental units of length and time. This will also be true of any other physical quantity defined in terms of force.

A relationship such as that between a force and the acceleration it produces when acting on a given mass (Fig. 3.3) in which a doubling of the force doubles the acceleration, a tripling of the force triples the acceleration and so on, is of a special type; the force and the acceleration are said to be *directly proportional* to one another. A relationship such as that between acceleration and mass for a constant applied force (Fig. 3.2) in which a doubling of the mass halves the acceleration, a tripling of the mass reduces the acceleration to one third of its previous value and so on, is of another special type; the acceleration and mass are said to be *inversely proportional* to one another.

Many different kinds of force exist. They include mechanical forces, produced by mechanical machines or engines of all kinds (including the forces produced by man or other living creatures), frictional forces, which always tend to bring motion to a halt, magnetic forces such as the force exerted by a magnet on a piece of iron, electric forces such as the force exerted by one electric charge on another, electromagnetic forces, which are a subtle mixture of electric and magnetic effects, and gravitational forces such as the attractive force exerted by the earth on anything near it. All these forces are defined quantitatively by Newton's second law of motion and the accelerations which they produce can be calculated using this law. Thus the motion of any body can be calculated if all the forces acting on it are known.

Force can also be defined in an equivalent way with the introduction of another useful quantity, the *momentum* of a moving body, which is defined as the product of the mass and the velocity of the body; the units of momentum are therefore kilogrammes multiplied by metres per second (kg m/s). Consider the motions produced in masses of 1 and 2 kg by a force of 1 N (lines marked F_1 in Fig. 3.3).

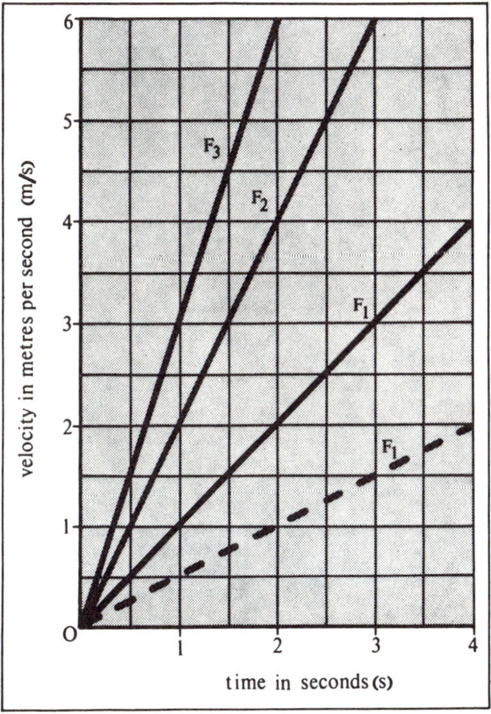

Fig. 3.3. Velocity-time graphs used in the definition of the magnitudes of forces of 1 N (F_1), 2 N (F_2) and 3 N (F_3). The solid lines represent the motions produced by each force acting on a 1 kg mass; the dashed line represents the motion produced by the 1 N force acting on a 2 kg mass. All masses are initially at rest.

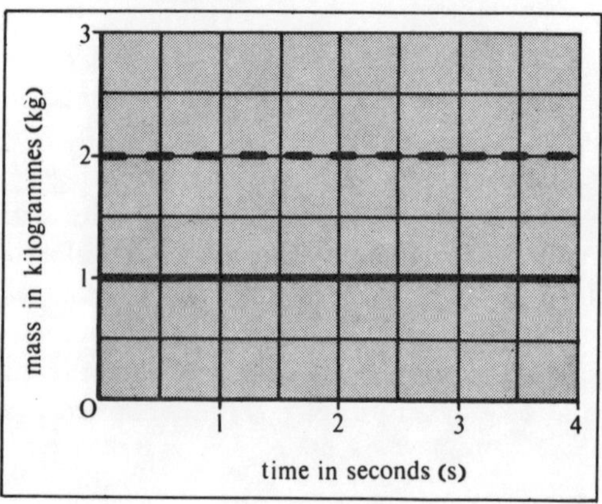

Fig. 3.4. Mass of the two bodies of Fig. 3.3 as a function of time.

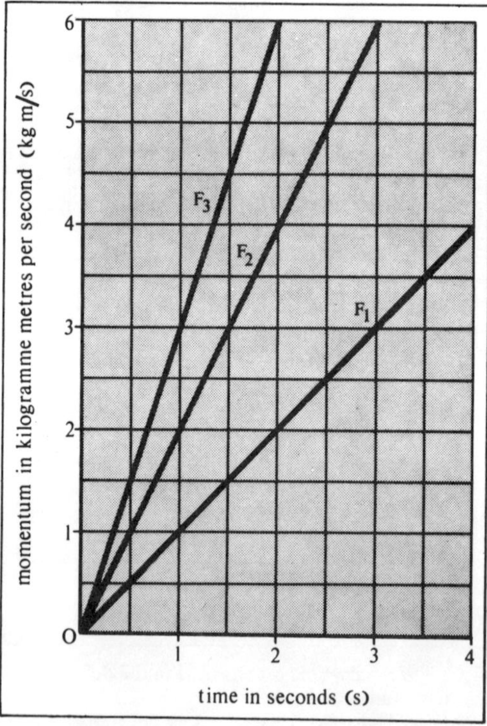

Fig. 3.5. Momentum-time graphs used in the alternative definition of the magnitudes of forces of 1 N (F_1), 2 N (F_2) and 3 N (F_3).

Suppose we now plot graphs showing how the masses of the two bodies vary with time when the force is applied (Fig. 3.4); since the masses are constant these graphs are horizontal straight lines. If we now multiply point-by-point the velocities of the masses at each point in time from Fig. 3.3 by the mass at each point in time from Fig. 3.4, we obtain the *momentum* at each point in time (Fig. 3.5). Note that both masses give the same straight line F_1 in Fig. 3.5 and that the slope of this line is 1 unit of *momentum* per second. The same momentum-time graph (line F_1) would be obtained for any mass which is acted on by a force of 1 N and in fact the force can be defined as the slope of the momentum-time graph or—what is the same thing—the rate of change of momentum with respect to time. This equality between force and rate of change of momentum with time is shown by lines F_2 and F_3 in Fig. 3.5 which are obtained by multiplying lines F_2 and F_3 in Fig. 3.3 by the 1 kg line in Fig. 3.4.

Another very useful attribute of momentum is that it is conserved in collisions. As a simple example of this, suppose we have a mass of 2 kg travelling at 3 m/s so that it collides with a mass of 1 kg which is at rest (Fig. 3.6). If the 2 kg mass continues in the same direction after the collision but with a velocity of 1 m/s, we can calculate the velocity of the 1 kg mass after the collision using the fact that momentum is conserved. The total momentum before the collision was 6 kg m/s (6 for the 2 kg mass and none for the 1 kg mass). After the collision the total momentum in the same direction is conserved

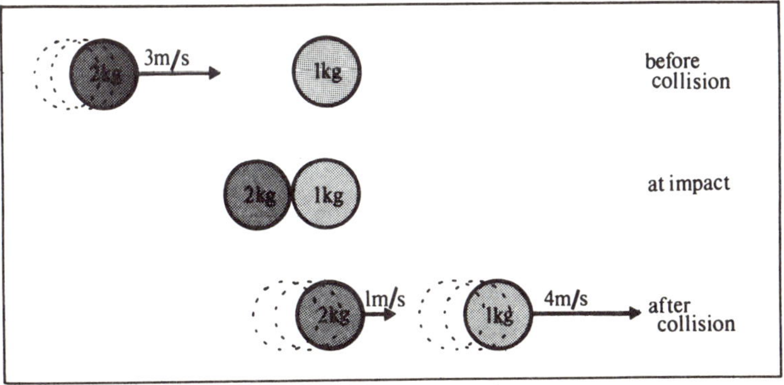

Fig. 3.6. Simple collision illustrating the conservation of momentum. The total momentum in the direction of the arrows is the same both before ((6+0) kg m/s) and after ((2+4) kg m/s) the collision.

and must still be 6 kg m/s but the 2 kg mass has only 2 kg m/s; therefore the 1 kg mass must have 4 kg m/s of momentum and hence must be travelling at 4 m/s. Similar calculations can be performed even if the masses reverse their directions of motion or move off at some angle to their initial directions after collision, or if more than two masses are involved. The principle of the conservation of momentum can be expressed by saying that the total momentum in any direction of a group of isolated bodies remains unchanged no matter what forces these bodies exert on one another either during direct collisions or in any other way. The principle can be derived fairly easily from the definition of force (or Newton's second law of motion) together with the additional fact that when any two bodies interact with one another they exert equal and opposite forces on one another (sometimes called Newton's third law of motion).

Two further physical quantities which are extremely important in dynamics and indeed the whole of physics are *work* and *energy*. *Work* is done when a force moves a body and is defined quantitatively as the product of the force and the distance moved by the body in the direction of the force. The units of *work* are therefore newtons multiplied by metres (Nm) and these units are usually called joules (J). Thus if a force of 2 N acts on a body and causes it to move a distance of 3 m in the direction of the force, then 6 J of *work* are done on the body. This precise definition of *work* obviously has some similarities with the intuitive or commonsense ideas of how hard one must work to push a heavy body from one place to another. *Energy* is a quantity which is defined as being equal to the amount of *work* which has been done on a body or which can be done in some way by a body.

As an example let us consider the case just mentioned above. If the body starts from rest and the 2 N force is the only force acting on the body, then, after moving 3 m, the body will have a finite velocity and is said to possess an *energy* of motion amounting to 6 J. Energy of motion is usually called *kinetic energy* and is only one of many forms that energy can take. *Potential energy* is another important form of energy which is related to the position of a body with respect to other bodies which are exerting forces on it. If we do work on a body by lifting it vertically against its own weight, that is against the gravitational attractive force exerted by the earth on the body, then the body acquires gravitational potential energy. The amount of such gravitational potential energy is equal to the gravitational force (or the weight of the body) multiplied by the vertical distance moved (ideally the body should be lifted so slowly that any kinetic energy it acquires is negligible). A body may similarly acquire magnetic, electric, electromagnetic or mechanical potential energies if work is done in moving the body against magnetic, electric, electromagnetic or mechanical forces.

All types of potential energy can be converted into an equal amount of kinetic energy under suitable conditions. For example, if 6 J of work were needed to lift a body vertically into a given position, it would then possess 6 J of gravitational potential energy. If the body were then allowed to fall vertically to its initial position it would lose its 6 J of potential energy but would gain 6 J of kinetic energy.

Energy can also exist in the form of heat, as was shown by the extensive experiments of Joule in the nineteenth century. In these experiments, measured amounts of mechanical work were done against frictional forces and the amounts of heat produced were always found to be directly proportional to the amounts of mechanical work done. Heat energy is also measured in joules as are all other forms of energy.

Energy is a very useful quantity since it has been established by experiment that, like momentum, it is conserved in an isolated group of bodies. The total energy of the isolated bodies remains constant no matter how the energy of the bodies is converted from one form to another or exchanged between them. Similarly if energy is added to an isolated group of bodies in some way, the total energy possessed by the group as a whole at any future time is always equal to the initial energy possessed by the group plus the added energy. And again this is true no matter how often energy may have been transferred from one body to another or may have been converted from one form into another. This attribute of energy is called the principle of conservation of energy which is sometimes expressed by saying that energy cannot be created or destroyed but only converted from one form to another.

Kinetic energy is particularly important in relativity theory and it will be useful to see how we can find the kinetic energy of a body in a simple example. Consider a body which starts from rest and is acted on by a constant force; its velocity will vary linearly with time as we have seen in Fig. 3.3. Let us find how the kinetic energy of this body varies with time. We can do this by first drawing Fig. 3.7(a) which shows how the position of or distance moved by the body varies with time. The curve in Fig. 3.7(a) is constructed from line F_1 in Fig. 3.3 so that its slope at any time is always equal to the velocity of the body at that time. Thus the slope at zero time is zero, while the slope after 2 s (tangent B of slope 2 m/s) is twice that after 1 s (tangent A of slope 1 m/s), the slope after 3 s is three times that after 1 s and so on. Fig. 3.7(a) is a special case of the accelerated motion shown in curve I of Fig. 1.8 in which the acceleration is constant so that the slope of the curve (velocity of the body) at any instant of time is proportional to the elapsed time. A curve of this form is called a *parabola*. Fig. 3.7(b) shows how the force acting on the body varies with time so that we can find the work done on the body, or its kinetic energy, by multiplying curves (a) and (b) point-by-point to give Fig. 3.7(c). Thus the kinetic energy curve as a function of time is also a parabola. Fig. 3.7 has been drawn for the special case of a force of 1 N acting on a mass of 1 kg. The same procedure, however, can be used to find the kinetic energy as a function of time for any value of force or mass, provided that the force remains constant.

The laws of dynamics together with the two general principles of conservation of momentum and energy were major successes of classical physics in that they accounted satisfactorily for a wide range of experimental observations and that for many years after they were established there was no significant experimental evidence against them.

All the laws and principles that have been discussed so far have been developed in terms of measurements which are implicitly assumed to have been made by an observer of some kind. And it should be clear that any other observer who is at rest with respect to the first observer could make exactly the same measurements in any experiment and thus would arrive at exactly the same laws and principles. If, however, the second observer is moving with respect to the first, then clearly some measurements, such as that of the velocity of a body moving with respect to the first observer, will be different for the two observers. As a simple example, if the second observer is actually attached to or sitting on the body which is moving with respect to the first observer, then the second observer will consider the body to be at rest with respect to him. The question

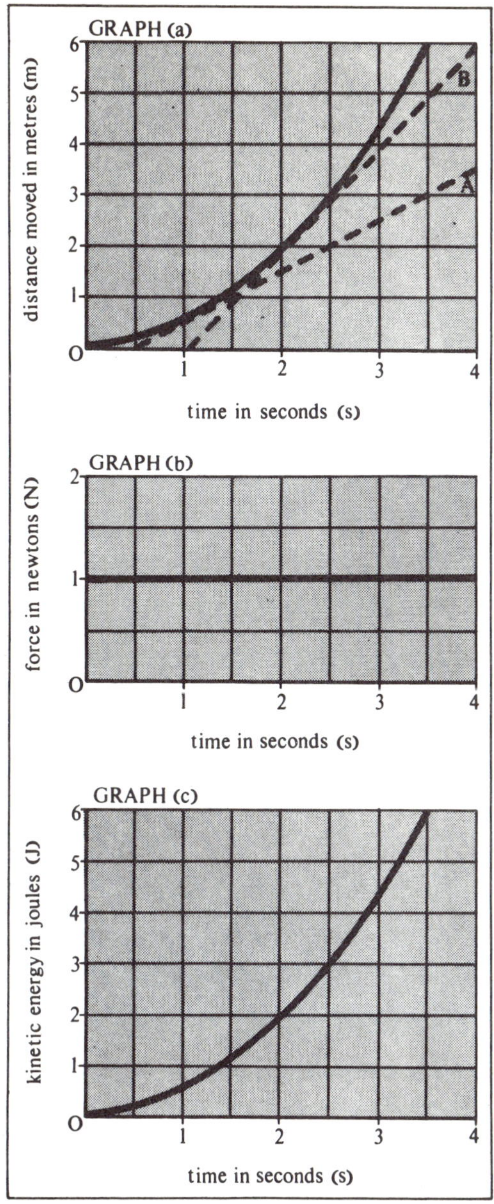

Fig. 3.7. (a) Distance moved as a function of time for a body of mass 1 kg, initially at rest, when subjected to a constant force of 1 N. (b) Force acting on the body as a function of time. (c) Kinetic energy of the body as a function of time. Graph (c) is obtained by the point-by-point multiplication of graphs (a) and (b).

as to how measurements and physical laws will appear to observers who are in relative motion with respect to one another is the central question of relativity theories and the answer given to it by classical physics is usually called the classical, Newtonian or Galilean principle of relativity. Classical relativity involves certain distinctions between different kinds of relative motion which are not immediately obvious and which are carried over into Einstein's work. We shall therefore now discuss classical relativity in more detail.

At first sight one might think that the likeliest way in which relative motion might affect physical laws would be that observers at rest would see the laws in one way and observers in any kind of relative motion would see them in different ways. This would be a seemingly sensible clear-cut distinction. In fact this is not the case. What is actually found is much more complicated. It is that there is a definite type of relative motion between observers, motion at a constant velocity (including zero velocity) with respect to each other, for which the physical laws perceived by each observer are the same. And for any other type of relative motion between observers (that is any motion involving a change in relative velocity or a relative acceleration), the observers perceive different physical laws.

The statements made in the last two sentences comprise the classical principle of relativity. As an example of this principle consider an observer inside a closed vehicle which is moving horizontally at a constant velocity with respect to the earth's surface. If he throws a ball vertically upwards, in his view it will rise and fall in a vertical straight line in just the same way as it would if he were at rest on the earth's surface. In fact identical results are obtained in any dynamical experiment carried out in the vehicle as would be obtained if the same experiment were carried out by an observer at rest on the earth's surface. This gives us an alternative way of stating the classical relativity principle: if we have two closed vehicles moving at a constant velocity with respect to each other, then it is impossible to detect or measure this relative motion by any experiment carried out inside either vehicle. It is only when one observer's vehicle is being accelerated or braked in some way that he sees any change in the up and down motion of the ball and the ball appears to behave differently for the two observers.

This distinction between different types of motion is the same distinction as was drawn in Newton's first law of motion, or the principle of inertia, and consequently observers moving with constant velocities with respect to each other are called *inertial observers* and observers in any kind of accelerated relative motion are called *non-inertial observers.* One might say that inertial observers are moving in a natural unforced way with respect to each other, while non-inertial observers are moving in a forced or non-natural way with respect to each other.

It is this fundamental distinction between relative motion with constant velocity and relative motion involving acceleration which is basic to both classical relativity and to Einstein's work, and on first acquaintance it is as surprising and as difficult to accept as any of Einstein's supposedly more difficult concepts. Yet it is there, right at the start of classical dynamics, in the work of Galileo and Newton. It seems strange to common sense since common sense has very little experience of carrying out experiments inside vehicles moving at constant velocities. Most vehicles move in ways that involve almost continuous changes in velocity which may be due to a variety of forces. Such forces usually include driving forces, frictional forces, forces due to winds or to bumps in the road or track, or even forces produced by a steering mechanism which change the direction of motion and thus also alter the velocity even if the speed is not changed. Common sense therefore feels that it can detect motion of any sort as it is familiar with the definite detectable effects inside the vehicle of all these forces.

The familiar vehicle whose motion most closely approximates to a constant velocity is the earth itself and, as we know, common sense does not easily accept that the earth is moving at all. Indeed when Copernicus first put forward his ideas concerning the motion of the earth, the common sense of the day (about four hundred years ago) asserted that the earth could not be moving as it could not be felt to be moving. And it is only by reference to bodies outside the earth, such as the sun and stars, that the almost constant velocity of the earth can be easily inferred. Experiments inside closed rooms on the earth's surface cannot detect the large, almost constant velocity of the earth with respect to the sun of about 30,000 m/s. However, careful experiments inside such rooms can detect the very small acceleration of about 0·03 m/s^2 due to the daily rotation of the earth about its own axis. And the acceleration of about 0·0006 m/s^2 due to the yearly rotation of the earth about the sun can also be detected inside such rooms by even more careful measurements.

The rest of this chapter on classical relativity and the next three chapters on Einstein's special relativity theory will deal with measurements and observations of the physical world carried out by observers whose relative motion is motion at constant velocity with respect to each other, that is with observations made by inertial observers. It is only in Chapter 7 that observers who are accelerated with respect to one another, that is non-inertial observers, will be considered both from the classical viewpoint and from that of Einstein's general relativity theory.

Although the classical relativity principle tells us that all inertial observers arrive at the same physical laws, this does not mean that they all make the same measurements of particular events. For example if an object is moving at 1 m/s with respect to a vehicle which is itself moving at 1 m/s with respect to the earth's surface and both these motions are in the same direction, then an observer on the vehicle will say the object is moving at 1 m/s while an observer on the ground will say it is moving at 2 m/s. However, they will both make the same measurement of the acceleration of any object. This is because measurements of accelerations only involve measurements of changes in velocities and not measurements of the absolute values of velocities and any change in the velocity of a body will be found to be the same by both observers. Therefore they will both observe Newton's second law of motion to be true.

In general the measurements made by two or more inertial observers can be compared most easily if each observer specifies the position in both space and time of any event observed by them all. Each observer can specify the position of the event in space in terms of three cartesian coordinates which refer to a certain set of axes at rest with respect to him. Each observer can also use a standard clock which is at rest with respect to him in order to specify the position of the event in time with reference to some arbitrary time origin or zero of time. This time is called the time coordinate of the event. The three space coordinates and the time coordinate make up a set of four co-ordinates which define the position of an event in space and time with respect to each observer. And it is obviously desirable and useful to be able to convert the coordinates of any event as measured by one observer into the equivalent coordinates of the same event as measured by any other observer. Such a conversion is called a *coordinate transformation* and the coordinate transformation which was derived in classical dynamics for a pair of inertial observers is a particularly simple one called the *Galilean transformation.* This transformation was obtained by what were considered to be obvious commonsense arguments. The time coordinates as measured by any two inertial observers were taken to be identical provided that the observers agreed on their origin or zero of time, while the space coordinates were transformed in a way similar to that used to transform the velocities in the example at the beginning of the last paragraph.

As a simple example of the Galilean transformation consider two inertial observers who are each situated at the origins of their own cartesian space coordinate axes at zero time. Let these axes coincide with one another at this time (see Fig. 3.8) and let the observers be

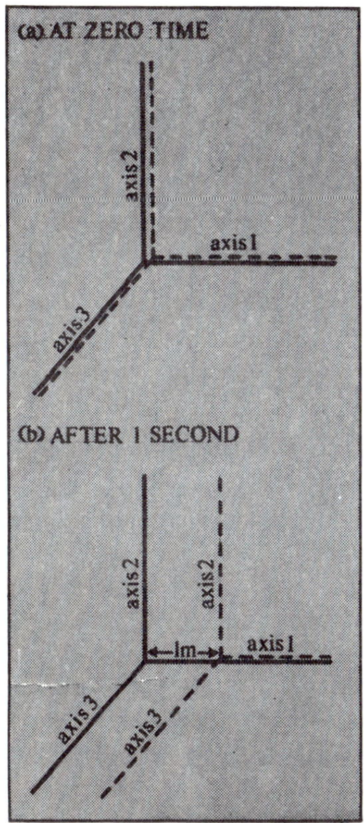

Fig. 3.8. Cartesian coordinate axes for two observers moving with respect to one another at a velocity of 1 m/s along axis 1.

moving at a velocity of 1 m/s with respect to each other along one of these axes (axis 1). An event which occurs at the same place as one of the observers (observer 1) after a time of 1 s will have a coordinate along axis 1 of zero according to observer 1 but will have a coordinate of 1 m along axis 1 according to the other observer (observer 2). The coordinates of the event along the other two axes will be zero for both observers so that the three position coordinates of the event will be (0, 0, 0) as measured by observer 1 and (1, 0, 0) as measured by observer 2. The time coordinate of the event is 1 s for both observers. This coordinate is sometimes included with the three space coordinates inside a pair of brackets, so that the full space and time coordinates of the event would be (0, 0, 0, 1) for observer 1 and (1, 0, 0, 1) for observer 2. If the event had occurred at the same place as observer 1 after 2 s, its coordinates would be (0, 0, 0, 2) for observer 1 and (2, 0, 0, 2) for observer 2. In a similar way the four coordinates (3 space and 1 time) of any event as measured by one inertial observer can be converted into four equivalent coordinates as measured by any other inertial observer. In these Galilean transformations the time coordinate of any event is always the same for any observer and it is only the space coordinates which change.

The simple examples discussed above of physical measurements made by two different inertial observers show that some physical quantities such as position coordinates or velocities are different for the different observers while others such as time coordinates are the same for all such observers. Quantities of the latter kind are said to be *invariant* for the two observers (or *invariant* under a Galilean transformation). Such quantities are extremely important since an invariant quantity can be considered to have some general objective usefulness since its value is independent of any particular observer. On the other hand *non-invariant* quantities may have some uses but their usefulness is more restricted since the values ascribed to them are different for each observer.

Since time coordinates are invariant, it is obvious that time intervals between events are also invariant under Galilean transformations and also that no problems arise in deciding on the time order of events or in deciding whether or not two events are simultaneous. Other invariants for inertial observers are distances (or space intervals) between any two events which occur at the same time and distances between any two points in space, since such distances are obviously independent of any coordinate system used to specify the positions of the events or points.

Accelerations will also be invariant under Galilean transformations since, as we have already seen, different inertial observers will describe the absolute velocities of bodies in different ways but will all perceive changes in velocities to have the same values. If we accept the assumption that the mass of a body does not vary with the velocity of an observer, then the invariance of acceleration implies that force is also an invariant. Also it is clear that any quantities formed by multiplying or dividing invariant quantities will also be invariant. However, quantities such as work, kinetic energy or momentum which involve non-invariant quantities like position coordinates or velocities will themselves be non-invariant.

Although the kinetic energy or momentum of an individual body is not invariant under a Galilean transformation, this does not necessarily mean that the principles of momentum and energy conservation are themselves invariant. In fact it turns out that the principles are invariant. Both momentum and energy are conserved in any closed system in the view of any inertial observer of that system. However, the actual numerical values ascribed by different observers to the total momentum in any direction or to the total energy of the system will be different. This means that these conservation principles retain their general usefulness and remain two of the major successes of classical physics.

The experimental evidence for the classical laws of motion, the conservation principles and the classical principle of relativity is enormous and has been built up over two or three hundred years of experimental work. Perhaps one of the major successes of classical physics was the way in which the motions of the planets around the sun could be explained in nearly every detail by combining together the laws of dynamics and Newton's expression for the gravitational force exerted by one body on another. Newton's law of gravitation gives the gravitational force exerted by one body on another solely in terms of the masses of the two bodies and the distance between the two bodies. The total gravitational force acting on any body is equal to the sum of the separate gravitational forces exerted on the body by all other bodies near enough to it to have a significant effect. (Incidentally this total force also turns out to be directly proportional to the mass of the body, a fact we shall refer to again in Chapter 7.)

A knowledge of this total force combined with the laws of dynamics enables the precise paths or orbits of the planets to be calculated. So precise had these calculations become that some unexplained irregularities in the orbit of the then outermost known planet Uranus led theoretical astronomers to suggest that the cause of the irregularities might be the gravitational force exerted by a hitherto unknown planet beyond Uranus. Their detailed calculations predicted the mass and precise orbit of the unknown planet and shortly afterwards a planet of this mass and orbit, now called Neptune, was discovered.

This discovery took place in 1846 and it was much later, in 1930, that a further planet, Pluto, was discovered by a similar method. One important discrepancy did remain in the theory of planetary motion. This was a discrepancy in the orbit of Mercury, the planet nearest the sun, and it was not satisfactorily explained by classical physics. This discrepancy will be mentioned again in Chapter 7. More recent evidence for the classical laws of motion has come from the extremely precise calculations which are essential for any successful space flight; all such calculations are detailed applications of the laws of motion which we have discussed.

One of the built-in assumptions of classical relativity, that mass is independent of velocity, has already been mentioned. We shall now discuss some of classical relativity's other implicit assumptions or limitations. However, we are only able to do this easily by the degree of hindsight provided by our knowledge of Einstein's work.

Possibly the most serious omission in the commonsense arguments leading to the Galilean transformations is the lack of any application of the operational viewpoint discussed in Chapter 2. In particular, this viewpoint is not applied in considering the measurements of time and position made by the various observers. If the operational viewpoint is applied to the question of how the clocks or time-scales of two inertial observers can be synchronized, then it is found that the idea of an absolute time which is the same for all observers can only be retained if time measurements from one clock can be transmitted instantaneously to an observer adjacent to any other clock. This question will be looked at in more detail in the next chapter. Here we shall just emphasize that the commonsense assumption that time is the same for all observers turns out to be equivalent to an assumption that information can be transmitted from one observer to another at an infinitely great speed—an assumption that does not appear to be an obvious common sense one at all.

In fact Einstein's criticisms of classical physics were principally directed at this assumption of the existence of an absolute time scale common to all inertial observers. We have seen that time measurement plays a fundamental role in the definition of nearly all the physical quantities involved in classical physics. Such quantities include velocity, acceleration, mass, force, momentum and energy, to name only a few. Hence it should be clear that if our time measurements are suspect, then the implications for classical physics will be serious.

This has been a long chapter but a very necessary one if we are to appreciate the background to Einstein's work and to have any quantitative grasp of the classical physics on which his work was based. Now, at last, we can turn to the special theory of relativity.

4
Special Relativity and Our Ideas of Space and Time

Einstein's special theory of relativity was first put forward in 1905 and since that time it has seen many developments and has been written about and interpreted in a variety of ways by many people. The fundamental ideas, however, were expressed very clearly by Einstein and so in this chapter his arguments will be used as a basis for the development of the main ideas of special relativity. The principles involved in the theory will be discussed first, the implications of these principles for our ideas on space and time will then be developed, and finally some of the experimental evidence for the new ideas will be given. This chapter is restricted to the impact of special relativity on kinematics, leaving to Chapter 5 the discussion of its effect on dynamics.

Special relativity, let us repeat and emphasize, is concerned with how physical laws appear to two or more observers who are moving with respect to one another with constant velocities. In other words the theory applies to inertial observers only and is a direct development of the classical relativity discussed in Chapter 3 which also deals only with inertial observers.

In fact the first general principle of special relativity is identical with the classical relativity principle and can be stated in either of the following exactly equivalent ways: 'Physical laws are the same for all inertial observers' or 'No experiment carried out by an inertial observer inside a closed vehicle can detect or measure his constant velocity with respect to any other inertial observer outside the vehicle'.

The second general principle was specifically introduced by Einstein when he insisted that all measurements or quantities which are used in special relativity theory, even the simplest and most obvious, should be defined operationally in the sense that this term was used in Chapter 2. We shall see how this can be done in a consistent way later in this chapter.

The third general principle, which was also introduced by Einstein, is that the speed of light (or of other electromagnetic radiations like light, such as radio or radar signals) is the same in any direction in empty space, is the same for all inertial observers and is independent of any motion of the body emitting the light.

The significance of the first two general principles should be clear from the discussions in Chapters 2 and 3, but the third principle requires some further explanation. First of all, one might ask, why is the speed of light or of other similar radiations involved in such a fundamental way in special relativity? The main reason for this is really a consequence of the second principle, the operational viewpoint. Some rapid method of conveying information from one observer to another is almost always required when consistent operational definitions are sought for quantities measured by different inertial observers.

The obvious way to convey such information is to use a light signal as this sort of signal is involved in every normal visual observation that we make and it is an experimental fact that light signals travelling in empty space provide the fastest method of transmitting information from place to place yet discovered. It is known that radiations similar to light such as radio or radar signals, or indeed any other electromagnetic radiation (see Appendix 2), travel in empty space at exactly the same speed of 3×10^8 m/s as light signals but no method is known of sending signals any faster than this. (In the rest of this book we shall refer indiscriminately to the speed in free space of light signals or of any other electromagnetic signals and shall regard such signals as entirely equivalent.) This is why the speed of light appears in the third general principle but we still have to justify experimentally the specific statements about the speed of light which the principle makes.

The direct experimental evidence for the independence of the speed of light of both the direction of motion of the light signal and the motion of the light source is reasonably strong. Careful laboratory experiments designed to detect any such dependence have made use of apparatus which can be rotated into any orientation with respect to the earth (and consequently with respect to the motion of the earth). These experiments which use a maximum source speed of about 30,000 m/s (the earth's speed with respect to the sun) have been repeated many times over the last eighty years or so and have shown no significant effects. More recent experiments using light signals emitted by beams of fast particles from the particle accelerator of the European Centre for Nuclear Research (CERN) at Geneva have shown that the speed of light is independent of the motion of the source for source speeds which are within one tenth of one per cent of the speed of light. However, probably the strongest evidence for the third principle is that so many of the predictions of special relativity, which are deduced using the principle, have been verified by countless experiments. Some of these experiments will be mentioned at the end of this chapter and of Chapter 5.

These basic general principles will now be used to develop operational definitions for the measurement of the positions of events in space and time. This will be done first for observers who are stationary with respect to each other and then for inertial observers.

Time measurements at or near any point in space can be made operationally by an observer at the point using a standard atomic clock of the type mentioned in Chapter 3. The first operational difficulty comes in relating the time measured at one point A in space (A time) with the time measured at another point B in space which is an appreciable distance from A (B time); that is, the difficulty lies in synchronizing the two identical atomic clocks at A and B and thus in deciding on the simultaneity or non-simultaneity of two events taking place at these points. This synchronization can be done as follows. Let the observer at A send a light signal to B and let the observer at B send a light signal back to A as soon as he observes the original signal from A. Observer A records the time on his clock at which he sent the original signal and also the time on his clock at which he receives the signal back from B. Since the light signal will travel at the same speed from A to B as from B to A (third general principle) and since the distance from A to B is the same throughout (A and B are stationary with respect to each other), we can be sure that the time

(A time) taken by the light signal in going from A to B will be the same as that taken in going from B to A. Therefore A will know that the signal is being received by B at an A-time which is exactly half-way between the A-time at which A sends the original signal and the A-time at which he receives the signal from B. If A and B both agree that the time at which B receives the signal is the zero time with respect to which all other A- and B-times are referred, then the clocks at A and B are operationally synchronized and there is no doubt as to whether events at A and B are simultaneous or not. A similar procedure can be used to synchronize the clocks used by any number of stationary observers situated anywhere in space.

Positions in space of objects or events can also be operationally defined by using a very similar method to the one just described for time measurements. Let observer A again send out a signal and receive back a signal from B but now observer B need not send back the signal himself. All he needs to do is to reflect back the original signal in some way. This is usually done in practice by using radio or radar signals, which are reflected by any material object, rather than light signals, although light signals could be used. Since A knows that the radar signal travels at the same speed in both directions, he knows that the distance from A to B is half the distance travelled by the signal in the time between his sending out the original signal and receiving back the reflected signal. For example if 4 s separate the sending of the original signal from receipt of the reflected signal, then the distance from A to B is the distance travelled by light in 2 s.

When distances are measured in this way (in terms of the time taken by a light signal to traverse the distance), it is usually more convenient to use a new unit of distance, the light-second (Ls), which is the distance travelled by light in 1 s. In our simple example the distance from A to B is 2 Ls. Distances in light-seconds (or light-minutes, light-hours or light-years) can be related to any other length standard by experiments in which the speed of light is measured in terms of the length standard. The use of the light-second as the unit of distance is very convenient in special relativity since then the speed of light is exactly 1 light-second per second (1 Ls/s) by definition and all other speeds are less than 1 Ls/s. We shall therefore usually measure distances in this way in the rest of the book.

Thus there are no difficulties in developing consistent operational definitions for the measurement of the positions of events in space and time for any number of observers provided that all the observers are at rest with respect to one another. The position of any event can be specified by measuring four coordinates (3 space and 1 time) and all observers at rest with respect to one another will agree on the values of these coordinates.

The next step is to extend these operational definitions to measurements of the four coordinates of any event as made by inertial observers. If, however, we continue to discuss special relativity in terms of four coordinates, it will be very difficult to give any explanations solely in graphical terms. Therefore in the rest of this chapter we shall only consider one dimension in space; that is, we shall consider the positions and motions of bodies which move along a single straight line so that only one position coordinate is needed for each event. Fortunately all the main results of special relativity can be illustrated using only one space dimension.

If we restrict ourselves in this way, it means that the position of any event in space and time can be specified by only two coordinates (1 space and 1 time) and so we can use diagrams drawn on a two-dimensional (2 space dimensions) page of this book to represent the events. Such a diagram, called a space and time or space-time diagram, is shown in Fig. 4.1 in which time coordinates, measured in seconds, are plotted horizontally and space coordinates, measured in light-seconds, are plotted vertically.

These diagrams are similar in many respects to the position-time graphs discussed in Chapter 1 (Fig. 1.7 and 1.8). Any point on the diagram represents an event with a definite space and time coordinate; for example the point A with coordinates $(1, \frac{1}{2})$ represents an event which took place 1 second in time and $\frac{1}{2}$ a light-second in distance away from the origin of the diagram. A stationary observer who is initially at the origin of both space and time axes will subsequently be represented by points along the time axis since his time coordinate changes while his space coordinate remains unchanged. Similarly a stationary observer who is initially 1 Ls from the origin of the space axis will subsequently be represented by points along the line I since once again his time coordinate changes while his space coordinate remains unchanged. In fact any stationary observer or object will be represented by a horizontal line on the space-time diagram (compare with line IV in Fig. 1.7).

Straight lines such as IIa or IIb, which have slopes of less than

1 Ls/s, represent observers or objects moving at constant speeds, less than that of light, with respect to the stationary observers mentioned above; IIa represents motion away from the position origin and IIb motion towards the position origin. The slope of each straight line gives the speed in question (compare lines I, II and III in Fig. 1.7). The observers represented by the time axis, and lines I, IIa and IIb are all inertial observers.

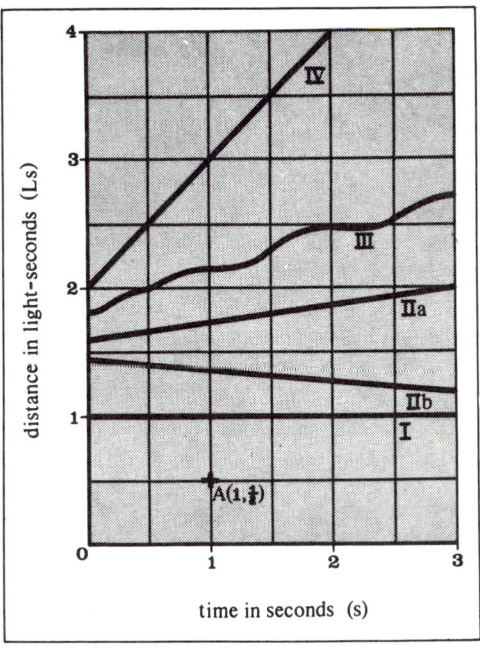

Fig. 4.1. Space-time diagram showing world-lines of inertial observers (lines I, IIa and IIb), a non-inertial observer (line III) and a light signal (line IV).

Curved lines such as line III represent observers or objects moving with variable velocities with respect to the stationary observers (compare lines I, II and III in Fig. 1.8); they can thus represent non-inertial observers. Straight lines such as line IV, which has a slope of 1 Ls/s, represent light-signals or objects moving at the velocity of light. Each point along any of the lines, I, IIa, IIb, III or IV on the space-time diagram represents an event in the 'life' of the object or observer concerned. The whole line, made up of all the points along it or all the events experienced by the object, represents the whole 'life' or history of the object or observer. These lines are called the *world-lines* of the object or observer and it should be clear that if any two world-lines intersect, the objects concerned will meet at the point in space and time represented by the intersection. As was mentioned earlier, no way is known of transmitting information (or indeed of transmitting any object which, of course, could always carry information) at speeds faster than that of light. Therefore all world-lines, straight or curved, must always have slopes which are less than 1 Ls/s.

The operational procedures, described earlier in this chapter, for synchronizing the clocks of two observers A and B who are stationary with respect to each other, and for measuring the distance between A and B, can easily be represented on a space-time diagram (see Fig. 4.2). The horizontal lines $A_1A_2A_3$ and $B_1B_2B_3$ are the world-lines of A and B respectively, while the line A_1B_2 is the world-line (of slope 1 Ls/s) of the emitted light signal and B_2A_3 is the world-line (also of slope 1 Ls/s) of the light signal reflected or transmitted from B. B_2A_3 is inclined downwards on the diagram because the

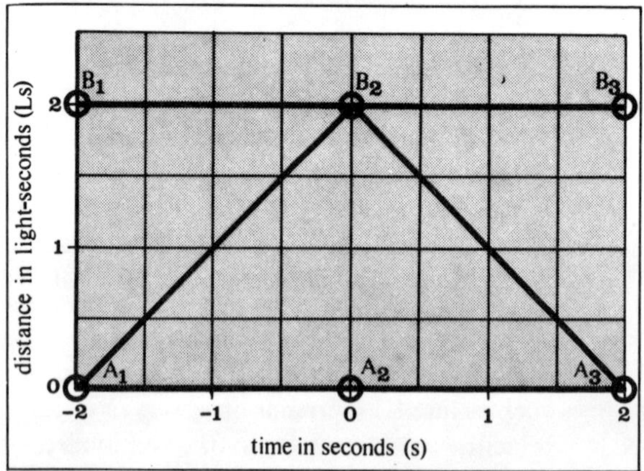

Fig. 4.2. Space-time diagram drawn by an observer A, showing the synchronization of the clocks of two observers, A and B, at rest with respect to each other. $A_1A_2A_3$ and $B_1B_2B_3$ are the world-lines of A and B respectively. A_1B_2 is the world-line of the emitted light signal and B_2A_3 is the world-line of the reflected light-signal.

reflected light signal is travelling in the opposite direction to the emitted signal. The point A_2, halfway in time between event A_1, the emission of the light signal by A, and event A_3, the detection of the returning signal by A, is the agreed time-zero which coincides in time with the event B_2, the detection of the emitted signal by B. Also the distance between A and B is equal to half the distance travelled by a light signal in the elapsed time A_1A_3 (4 s); that is, it is 2 Ls.

As was mentioned earlier, we can use a similar procedure to synchronize any number of clocks which are all stationary with respect to A and B and which are situated at any distance from A and B. Thus we can unambiguously define a time scale with respect to a whole family of observers situated anywhere in space provided that they are all stationary with respect to one another; let us call this time scale A-time.

Now let us introduce another observer C who is moving at a constant velocity with respect to A and B (and the whole family of observers mentioned above) and let us view everything from the point of view of C. Observer C can draw his own space-time diagram (Fig. 4.3) and, by using procedures exactly like those described above, he can operationally define the position and time coordinates of all events as he observes them. In particular he can synchronize any number of clocks, all stationary with respect to him, situated at any point in space, and so unambiguously define a time-scale for another whole family of observers situated anywhere in space provided that they are all stationary with respect to C; let us call this time-scale C-time.

Now let us consider how C will represent on Fig. 4.3 the sequence of events $A_1B_2A_3$ shown from A's point of view in Fig. 4.2. The world-lines of A and B will now be inclined parallel straight lines as shown, since A and B are both moving at the same constant velocity with respect to C (note that it has not been assumed in drawing these world-lines in Fig. 4.3 that the distance A_1B_1 between A and B as measured by C is the same as the distance A_1B_1 as measured by A). However, the world-lines representing the light signals emitted by A and reflected by B will still have slopes of 1 Ls/s in C's space-time diagram just as they did in A's diagram. This fact is a consequence of that part of the third general principle of special relativity which states that the speed of light is the same for all inertial observers. Therefore the emitted light signal will be represented on Fig. 4.3 by the line of slope 1 Ls/s which is shown starting from A_1. The intersection of this line with the world-line of B will locate the event B_2 as observed by C. Similarly the reflected light signal will be represented on Fig. 4.3 by the line of slope 1 Ls/s which is shown starting from B_2 (once again, this line is inclined downwards since the reflected signal is travelling in the opposite direction to the emitted signal). The intersection of this line with the world-line of A will locate the event A_3 as observed by C. Thus the emitted and reflected light signals will be represented on Fig. 4.3 by the world-lines A_1B_2 and B_2A_3.

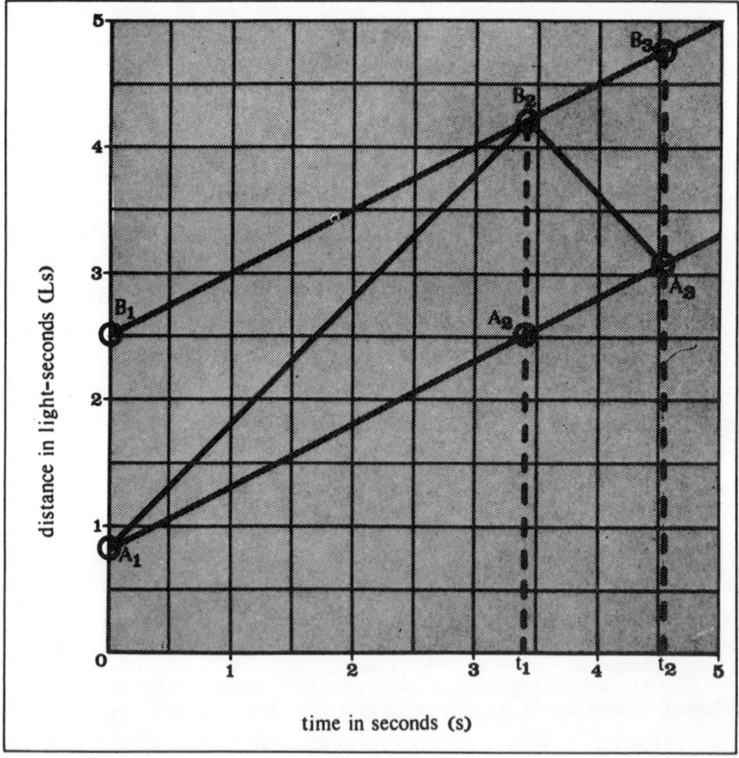

Fig. 4.3. Space-time diagram showing the same events as Fig. 4.2 as seen by a third inertial observer C moving at a constant velocity with respect to A and B. $A_1A_2A_3$ and $B_1B_2B_3$ are the world-lines of A and B respectively. A_1B_2 is the world-line of the emitted light-signal and B_2A_3 is the world-line of the reflected light signal.

Finally, let us consider how C records the C-time t_1 of the event B_2. From Fig. 4.3 it is clear that t_1 is not half-way in C-time between the events A_1 (C-time zero) and A_3 (C-time t_2). According to A, however, the event B_2 lies exactly half-way in A-time between the events A_1 and A_3 (see Fig. 4.2). It is clear that the observers A and C disagree about the time coordinate of the event B_2 and therefore we have the extraordinary situation from the classical point of view that two inertial observers (A and C) who both operationally establish perfectly consistent time scales (A-time and C-time) will differ in the time coordinate that they assign to the same event B_2.

The absolute time scale, common to all inertial observers, which was assumed to exist by Galileo and Newton is no longer valid: the operational viewpoint combined with the constancy of the speed of light have destroyed it. The consequences of this fact for classical physics are obviously serious, since so many quantities in classical physics were defined in ways which directly or indirectly involved time measurements. A glance at Fig. 4.3 shows that the discrepancy between the A-time and the C-time of the event B_2 will become greater as the velocity of A with respect to C increases, since then the slope of the world-lines of A and B will also increase. On the other hand if the velocity of A with respect to C is very small compared with the speed of light, the world-lines of A and B on Fig. 4.3 are very nearly horizontal; under these conditions A and C will be in agreement over the time of B_2—it will be half-way between A_1 and A_3 in both A-time and C-time.

Thus the time-scales of A and C (A-time and C-time) will coincide for small relative velocities of A and C so that the ideas of special relativity are asymptotic to the classical ideas (see Chapter 2) for relative velocities small compared with the speed of light. Alternatively we can say that absolute time is valid, provided that the speed of light is much greater than the speed of any other object which is being observed; under these conditions we can regard light signals as travelling instantaneously from point to point in space (see Chapter 3).

Note that the arguments of the last three paragraphs are valid no matter what assumption is made about the distance between A and B as measured by C (A_1B_1 on Fig. 4.3) compared with the same distance as measured by A (A_1B_1 on Fig. 4.2). The crucial step in showing that A and C assign different time coordinates to the same event is that the world-lines of the light signals have the same slopes in the two diagrams. In other words the ambiguity in the time coordinates of the event is a direct consequence of one of the basic principles of special relativity.

This ambiguity in time measurements has the further serious consequence that measurements of distances, and hence of position coordinates in space, will also be different for different inertial observers. This occurs because our operational procedure for distance measurements (see earlier in this chapter) makes direct use of time measurements. Once Einstein realized that both the space and time coordinates of an event were different for different inertial observers, he next tried to obtain a new coordinate transformation which was consistent with the principles of special relativity and which would replace the Galilean transformation discussed in Chapter 3.

The detailed arguments he used are too complex to be set out in a book of this kind but the main steps involved are not too difficult to outline. The desired coordinate transformation would tell us how to convert the position and time coordinates of any event as measured by one inertial observer into the position and time coordinates of the same event as measured by another inertial observer moving at any constant velocity v Ls/s with respect to the first observer. The key steps in Einstein's argument were that a single event as viewed by the first observer should correspond to a single event as viewed by the second observer, that the speed of light should be the same for both observers, and that the transformation should be reversible (that is, if we take a transformed set of coordinates for an event and apply the transformation again, this time with respect to an observer moving at a constant velocity of the same magnitude but opposite in direction to v, we should get back to exactly the same coordinates that we started with, provided that we choose the coordinate origins carefully).

The resulting transformation was more complicated than the Galilean transformation in that both the position *and* the time coordinates of an event as viewed by the second observer were each found to depend on the position and time coordinates of the first observer and on the velocity v; in the Galilean transformation it was only the position coordinate which showed such a complicated dependence.

The most serious implication of this new transformation, usually called the Lorentz coordinate transformation (see Chapter 6), is that space and time can no longer be regarded as separate concepts, independent of each other and of any observer. We must rather speak of a composite entity called space-time which each inertial observer breaks down into space and time in his own way and thus obtains his own space-time coordinates (which differ from the space-time coordinates of most other inertial observers) for any events which he observes.

The Lorentz coordinate transformation gives exactly the same results as the Galilean transformation if v is a lot less than 1 Ls/s (the speed of light) so that the asymptotic behaviour of the two transformations is as it should be. If, however, v is not very much less than 1 Ls/s, then the Lorentz transformation predicts some effects which appear strange to common sense. One such effect, usually called time dilation, is the prediction that the time between two events like A_1 and A_3 in Fig. 4.3 will be greater as measured by C than as measured by A; that is, as far as the stationary observer C is concerned, the clock of the moving observer A appears to be going more slowly than C's own clock and thus time as measured by A appears to be dilated compared with time as measured by C. The Lorentz transformation enables us to calculate precisely the amount of time dilation to be expected for any value of the relative velocity of the two observers.

This effect of time dilation provides one example of a quantity, time interval between two events, which was invariant for a Galilean transformation but is not invariant for a Lorentz transformation. Another Galilean invariant, the length of an object as measured by inertial observers who are stationary or moving with respect to the object, is also found not to be invariant for a Lorentz transformation. In fact the length of the object is measured to be less if the object is moving with respect to the observer than if the object and the observer are stationary with respect to one another; this effect is usually called a length contraction.

Invariant quantities, however, were very useful in classical physics. Therefore it was natural to try and find quantities which would be invariant in special relativity (for a Lorentz transformation) in the same way as time intervals or lengths were invariant in classical relativity (for a Galilean transformation). There is no systematic way of finding such invariants; rather it is a matter of trial and error, inspired guesswork or intuition. The simplest invariant in special relativity is a quantity called the *interval* between two events which is defined in terms of both the position and time coordinates of the events. In defining *interval* we shall again restrict ourselves to one space dimension so that, as before, the position of any event in space and time can be specified by one time coordinate, which we shall measure in seconds, and one space coordinate, which we shall measure in light-seconds. The definition can easily be extended to three space dimensions but it is not then possible to draw simple space-time diagrams.

Consider two events, one of which occurs at the origin of the space and time coordinates used by one particular inertial observer (event O in Fig. 4.4 with time and position coordinates (0, 0)) and the other occurs at some other point in this observer's space-time (say, event C in Fig. 4.4 with time and position coordinates (2, 1)). The *interval* between these two events is defined in a way that recalls the method of Pythagoras for finding the distance between two points in space (Chapter 3). The square of the time coordinate of C is found (2^2 or 4) and then the square of the position coordinate of C is also found (1^2 or 1). Then the square of the position coordinate is subtracted from the square of the time coordinate and the result is defined to be the square of the *interval* between the two events O and C (4 minus 1 is 3 which is the square of the interval between O and C; the symbol s is often used for interval so in this case we can write $s^2=2^2-1^2=4-1=3$).

This may seem a somewhat arbitrary or abstract way of defining a physical quantity but it is useful because it is invariant under a Lorentz transformation; that is, if we find the coordinates of the same two events as measured by any other inertial observer by applying a Lorentz transformation to the coordinates (0, 0) and (2, 1) and then calculate the interval between the events in the same way as before but using the new coordinates, we find that the square of the interval, s^2, will still be 3 although the individual position and time coordinates of the events will in general be different. So this new invariant quantity of special relativity is a peculiar mixture of classical space and time with some resemblances to the old invariants and some striking differences; the composite name space-time for the coordinate framework used in special relativity now seems even more appropriate.

Some further attributes of intervals between events will now be discussed with the aid of Fig. 4.4 which is the space-time diagram for an inertial observer. The points O, A, B, C, D and E represent events whose coordinates as measured by this observer are shown in brackets. The line OL has a slope of 1 Ls/s and represents the world-line of a light signal emitted by the observer as he passes through the point O in space-time (compare line IV in Fig. 4.1). We have already calculated the square, s^2, of the interval, s, between events O and C to be 3; let us write this $s^2_{OC}=3$. Now let us calculate the squares of the intervals between O and the other events shown in Fig. 4.4. If we do this we find $s^2_{OA}=1^2—0^2=1$, $s^2_{OB}=0^2—1^2=—1$, $s^2_{OD}=1^2—2^2=—3$ and $s^2_{OE}=1^2—1^2=0$. The values of s^2, the square of the interval, can be either positive (s^2_{OA}, s^2_{OC}), negative (s^2_{OB}, s^2_{OD}) or zero (s^2_{OE}) and these three possibilities correspond to three different types of interval. Intervals with s^2 positive are called *time-like intervals* since they can be classed with the interval OA which

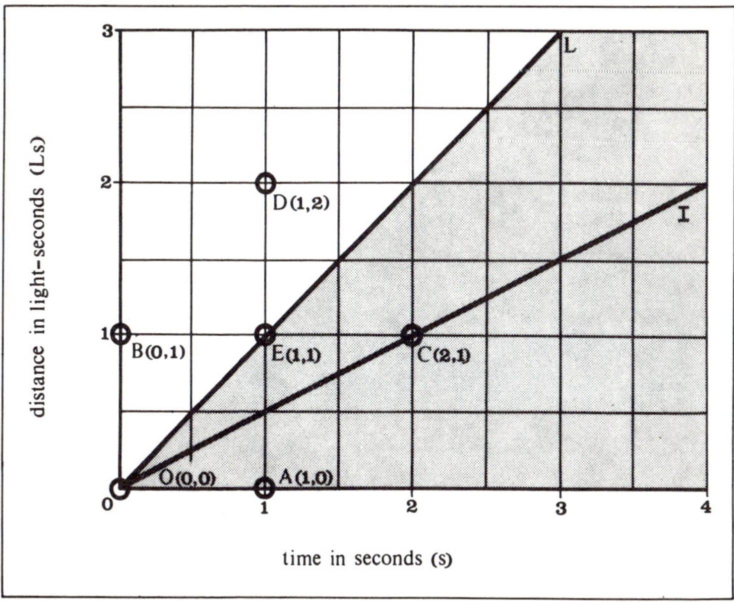

Fig. 4.4. Space-time diagram used to illustrate the properties of intervals in space-time. Line I is the world-line of an inertial observer while OL is the world-line of a light signal.

involves two events that have the same space coordinate and that only differ in their time coordinates. Intervals with s^2 negative are called *space-like intervals* since they can be classed with the interval OB which involves two events with the same time coordinate but different space coordinates. Intervals with s^2 zero are called *null intervals.*

It should be clear that the interval between O and any event like E lying along OL will be a null interval since all events along OL will have equal values for their time and position coordinates. Similarly the interval between O and any event in the shaded region of Fig. 4.4 will be time-like since all the events in this region have time coordinates which are greater than their space coordinates, while the interval between O and any event in the unshaded region will be space-like since all the events in the unshaded region will have space coordinates which are greater than their time coordinates. Thus the world-line, OL, of a light signal divides space-time into two distinct regions and the difference between these regions turns out to have a definite physical significance which we shall now explore.

It is always possible for a world-line (such as line I in Fig. 4.4) to be drawn between any two events separated by a time-like interval (such as O and C in Fig. 4.4). On the other hand no world-line can be drawn between any two events separated by a space-like interval (such as O and B in Fig. 4.4) since such a world-line would need to have a slope greater than 1 Ls/s. This would mean that the object or observer represented by the world-line would be travelling at a speed greater than that of light which contradicts the third principle of special relativity. Thus events lying in the unshaded region of Fig. 4.4 cannot be reached from event O by any observer or object who was also at the point O in space-time, whereas events in the shaded region can be reached from O by such an observer or object.

Now consider an inertial observer moving along a world-line (line I) between events O and C. The space coordinates that he measures for both O and C will be zero (compare the space coordinates of events O and A as measured by the original inertial observer of Fig. 4.4), and so the interval between O and C, s_{OC}, will be equal to the time that he measures between events O and C. Since the time recorded by such an observer is equal to the invariant interval between the events concerned, it is sometimes called the *proper time* between the events. This name, however, is a little misleading since it seems to imply that there is something improper or disreputable about the measurements made by any other observer of the times of the two events.

Finally consider the limiting case of an inertial observer moving at the speed of light along the world line OL between events O and E. He will record zero time for both events since s_{OE} is zero (the interval is a null interval). One might consider this to be the ultimate in time dilation!

Since the Lorentz transformation enables the position and time coordinates of any event as measured by an inertial observer to be transformed into the equivalent coordinates of the same event as measured by any other inertial observer, it is obviously possible to derive similar transformations for any quantities which are derived solely from measurements of position and time coordinates. Thus transformations can be found for velocities and accelerations, and for any other kinematic quantities. These transformations are a little more complicated mathematically than the Lorentz transformation but involve no new points of principle. Therefore we shall not discuss them in detail except to mention one point concerning the velocity transformation.

We have already mentioned a simple example of the velocity transformation of classical relativity in introducing the Galilean coordinate transformation in Chapter 3. There we said that if an object which is inside a vehicle moving on the earth's surface at 1 m/s, has a velocity of 1 m/s with respect to the vehicle, and if the two velocities are in the same direction, then the object has a velocity of 2 m/s with respect to the earth's surface; that is, we just added the two velocities in this simple case.

In classical relativity it is assumed that a similar simple addition of velocities is valid no matter how large the velocities are. The velocity transformation of special relativity does give the same result, 2 m/s (within the limits of accuracy to which we can make the measurement) for velocities like these which are extremely small compared with the velocity of light. However, for velocities which are comparable with that of light, the new transformation gives different results. In particular if both velocities involved are equal or nearly equal to the velocity of light (3×10^8 m/s), then the resulting velocity is not 6×10^8 m/s (classical transformation) but is still

3×10^8 m/s (special relativity transformation). This result is therefore in agreement with the third general principle of special relativity and with the CERN experiment on the velocity of light signals emitted by fast moving sources which was mentioned earlier in this chapter. The new velocity transformation therefore has two desirable properties—it is asymptotic to the old transformation and it is in agreement with experiment.

Most of this chapter has necessarily been theoretical. Even though the experimental emphasis has been present in the operational viewpoint, the experiments mentioned in operational definitions are usually imaginary or thought experiments, capable of being performed but not usually actually carried out. Now let us look at two experiments which provide direct evidence for some of the ideas of special relativity which have been discussed in this chapter.

The first experiment is an attempt to see if electrically charged particles could be accelerated to speeds greater than that of light. The particles used were electrons which are the lightest type of charged particle and therefore the easiest to accelerate to high speeds. Pulses or bursts of electrons were accelerated to high speeds by means of large electrical voltages (up to 15 million volts) and then the speed of each accelerated pulse was measured directly by timing the pulse over a measured distance. The distance used was about 10 m and the times involved were sometimes as short as one ten millionth of a second or less. Modern electronic methods, however, are sufficiently precise for such short times to be measured with reasonable accuracy.

Line I in Fig. 4.5 shows the relationship between the speed of the electron pulse in Ls/s and the accelerating voltage which would be obtained if classical dynamics were correct; the electrons should reach the speed of light at an accelerating voltage of about 250,000 volts and should reach even higher speeds at higher voltages. Line II in Fig. 4.5 is a smooth curve drawn through the measured data points (some of which are shown by crosses in the figure). Thus Line II shows the speeds actually measured for different accelerating voltages. It is clear that the two lines are asymptotic for voltages below about 50,000 volts and that for higher voltages the electron speed is not given by Line I but tends to a limiting value equal to 1 Ls/s, the speed of light.

The maximum voltage used in the experiment (not shown in Fig. 4.5) would have produced a speed of eight times the speed of light if

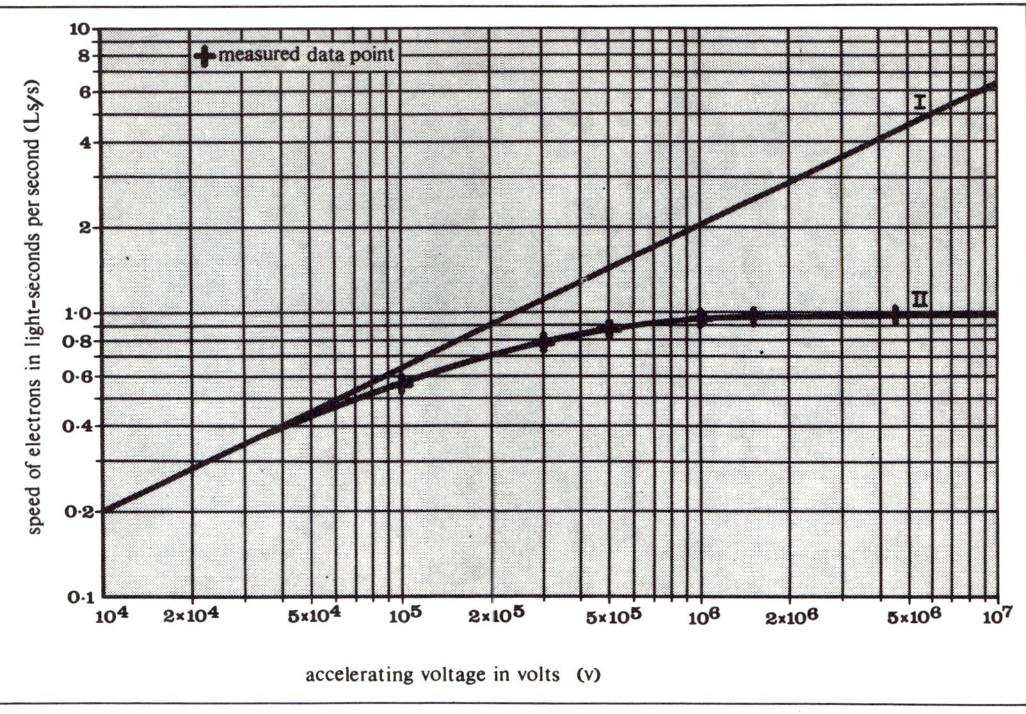

Fig. 4.5. Speed of electrons as a function of accelerating voltage. Lines I and II show respectively the relationships predicted by classical and special relativistic dynamics.

classical dynamics were correct. This experiment provides direct evidence that classical dynamics is incorrect for fast-moving particles and also that the speed of light seems to be a limiting speed which cannot be exceeded by moving particles. It therefore supports the third basic principle of special relativity.

The second experiment is a direct observation of time dilation. The clocks used in this experiment were rather unusual. They consisted of beams of radioactive particles produced by the same CERN particle accelerator which we have already mentioned. The particular particles used in this experiment (called π-mesons or pions) are well known in nuclear physics and have been observed in many experiments. They are known to be unstable and to change or decay into some other type of particle in an average time of about three hundredths of a millionth of a second (3×10^{-8}s) when they are at rest. An average lifetime of this size would mean that pions could not travel a distance of more than about 10 m even if they were moving at the limiting speed, the speed of light (3×10^8 m/s). The pions produced by the CERN accelerator are travelling nearly at this limiting speed with respect to observers stationary on the earth's surface and are found to travel for an average distance of about 250 m before decaying. Thus the average lifetime of the pions as observed by such observers has apparently been extended by a factor of about 25. Alternatively we can say that the time-scale of the pion has been slowed by this factor or that its time-scale is dilated compared with that of the observer on the earth's surface.

The amount of this dilation is in very good agreement with that predicted by the Lorentz transformation, so that the experiment provides both qualitative and quantitative evidence for the correctness of special relativity. In fact if time dilation did not occur, many of the successful experiments carried out at CERN and at other similar laboratories would not be possible since the particles produced by the accelerators would not exist long enough for them to reach the particle-detecting apparatus. Similar results have been observed with fast-moving unstable particles produced at heights of about 60 km above the earth by energetic cosmic rays. If time dilation did not occur, such particles would decay before they had travelled more than about 1 km, but in fact large numbers of them are observed at sea level.

Further indirect evidence in favour of special relativity will be discussed at the end of the next chapter.

The three basic principles of special relativity have thus led us to a totally new attitude to space and time. We have also seen that the experimental evidence is in favour of the results of special relativity. The ground we have covered in this chapter has therefore shown that the foundations of classical physics and classical relativity are very uncertain. In the next chapter we shall see how classical physics, and dynamics in particular, has to be modified so as to be consistent with our new ideas of space and time.

5
Special Relativity and Dynamics

The two fundamental principles of classical dynamics are the definition of the natural motion of a body as being motion with constant velocity (Newton's first law) and the quantitative definition of force in terms of mass and acceleration (Newton's second law). All the other physical quantities and laws which were introduced in Chapter 3 are developments of these two principles.

The first of these fundamental principles is completely retained in special relativity. The natural motion of a body on which no forces are acting is still taken to be motion with constant velocity. Therefore the world-line of a naturally moving body on a space-time diagram will be a straight line like the straight lines II(a) and II(b) in Fig. 4.1. Different inertial observers will assign different velocities to the motion of such a body and so will represent its motion by straight world-lines of different slopes, but all inertial observers will assign straight world-lines to freely or naturally moving bodies.

This retention of the principle of inertia or Newton's first law of motion in special relativity is really implicit in the first principle of special relativity which singles out inertial observers from all other observers in just the same way as does classical relativity.

Newton's second law of motion, however, is not retained intact in special relativity so that the rest of dynamics is modified too. This chapter explains how the main modifications were made.

The three basic principles of special relativity that led to the new ideas of space and time described in the last chapter do not give any direct way of finding new and more general dynamical laws to supersede the classical ones. The new laws were discovered by the use of a certain amount of intuition (or guessing), and the test of their correctness is, as always, the experimental one. There are, however, some general guidelines which were useful in this process. First, the new laws must be asymptotic to the classical laws for speeds that are small compared with the speed of light. Secondly, there should be quantities in the new theory which are invariant under a Lorentz transformation. Thirdly, it would be highly desirable if useful generalizations similar to the principles of momentum and energy conservation were to form part of the new laws. In fact it is by considering ways of generalizing the law of conservation of momentum that new dynamical laws consistent with the principles of special relativity can be arrived at most simply.

Consider again the simple collision illustrated in Fig. 3.6 which was used in Chapter 3 to illustrate the conservation of momentum. Fig. 3.6 views the collision from the point of view of an observer who is stationary with respect to the 1 kg mass before the collision, since this mass was then at rest. If this same collision is viewed by another inertial observer, for example an observer who is stationary with respect to the 2 kg mass after the collision, then it is easy to calculate the velocities of all the masses both before and after the collision using the Galilean velocity transformation. All that is needed is to subtract 1 m/s (the velocity of the 2 kg mass) from all the velocities and so obtain the values shown in Fig. 5.1. It is clear that momentum

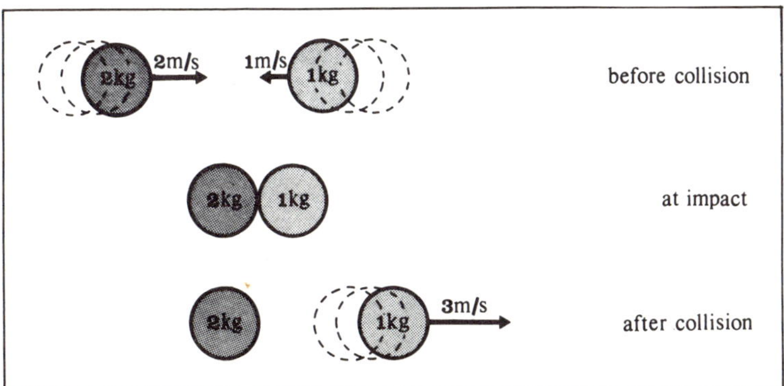

Fig. 5.1. The simple collision of Fig. 3.6. as viewed by another inertial observer. The total momentum from left to right is still the same before ((4−1) kg m/s) and after ((0+3) kg m/s) the collision.

is still conserved in the collision although the total momentum is not invariant (a total of 6 kg m/s when viewed by the first observer but only 3 kg m/s when viewed by the second observer).

If, however, the velocity transformation of special relativity is used instead of the classical Galilean transformation to see how the second observer views the collision, then the transformed velocities obtained will be different from those shown in Fig. 5.1 and the second observer will find that momentum is not conserved in the collision. This sort of calculation can be performed for any collision and it is always found that momentum, defined as the product of mass and velocity (see Chapter 3), is not conserved if the velocity transformation of special relativity is used. In these momentum calculations, however, we have assumed that any observer will measure the same value for the mass of a body no matter how fast the body is travelling with respect to him. In other words, we have assumed that the mass of a body is invariant and is always equal to its rest mass (see Chapter 2). Let us now drop this assumption and make another one instead. Let us assume that the mass of a body (as observed by any observer) varies with its speed (as measured by the same observer) in the way shown by curve II of Fig. 2.1. If we make this new assumption about mass and combine it with the velocities given by the velocity transformation of special relativity, then we find the surprising result that momentum, still defined as the product of mass and velocity, is always conserved in all collisions. This result suggested very strongly that the momentum of a body should be redefined in special relativistic dynamics as the product of its velocity with a mass which is varying with speed in the manner shown by curve II of Fig. 2.1. Such a redefinition of momentum means that the very useful principle of momentum conservation is retained in special relativistic dynamics. Also the principle is retained in a form which is asymptotic to the classical principle since curve II in Fig. 2.1 coincides with curve I for speeds which are small compared with the speed of light, and we also know that the velocity transformations of classical and special relativity are equivalent for small velocities.

This first tentative step towards the establishment of special relativistic dynamics obviously depends on the correctness or otherwise of the assumed variation of mass with speed. Fortunately it was not long (1908) before some fairly direct experimental evidence was discovered which supported the assumed variation. And once again the experiments used those very light electrically charged particles, electrons. Electrons with speeds of up to about 0·7 Ls/s were subjected to electric and magnetic forces and their subsequent motions were observed. From these observations it was possible to calculate a value for the electric charge of each electron divided by its mass. The results are shown in Fig. 5.2(a) and it is clear that this ratio of charge to mass is decreasing as the speed of the electrons increases. If we assume that the electric charge of an electron does not change with speed (see later for evidence for this assumption), then the results show that the mass of the electrons must be increasing as their speeds increase and approach 1 Ls/s, the speed of light. It is not immediately obvious, however, whether the mass of the electrons is increasing with speed precisely in the way shown by curve II in Fig. 2.1; for easy reference, this curve has been redrawn in Fig. 5.2(b) with the same scale along the speed axis as is used in Fig. 5.2(a). We can use Fig. 5.2(a) and 5.2(b) to see whether the mass increase is equal to that assumed in the last paragraph in the following way. We can read off the ratio of charge to mass at any speed from Fig. 5.2(a) (e.g. this ratio is 1·41 arbitrary units at 0·6 Ls/s), and then read off the assumed ratio of mass to rest mass at the same speed from Fig. 5.2(b) (1·25 at 0·6 Ls/s). If we then multiply these two ratios we should obtain the ratio of the charge of an electron to its rest mass $\left(\frac{\text{charge}}{\text{mass}} \times \frac{\text{mass}}{\text{rest mass}} = \frac{\text{charge}}{\text{rest mass}}\right)$ if the assumed mass variation is correct. In our example the product of the two ratios is (1·41 × 1·25) or 1·76 arbitrary units at 0·6 Ls/s, while the ratio of charge to rest mass for an electron is also 1·76 arbitrary units, as can easily be seen from Fig. 5.2(a). Thus the assumed mass variation agrees with experiment at 0·6 Ls/s. In a similar way we can check the assumed mass variation at any other speed by multiplying point-by-point the curves in Fig. 5.2(a) and 5.2(b); this has been done in Fig. 5.2(c). The result is always 1·76 arbitrary units, within the limits of accuracy of the measurements, and so the variation of mass with speed suggested by special relativity is confirmed by these experimental results.

This experimental support for the mass variation made it plausible for a new special relativistic definition of force to be made. This was done by analogy with the alternative classical way of defining force as the slope of the momentum-time graph, or the rate of change of the momentum of a body with time, which was discussed in Chapter 3. In special relativity force is also defined as the rate of change of the momentum of a body with time, but now the momentum is calculated in the new way mentioned above; that is, the momentum of a body is the product of its velocity and a mass which is not constant but is varying with speed in the way shown in Fig. 5.2(b). This new definition of force is obviously asymptotic to the old one and can be regarded as a generalization of Newton's second law of motion which is compatible with the principles of special relativity. This new generalized law can be applied in a similar way to Newton's classical law to calculate the change in the motion of any body due to any force acting on it; however, since the new law is more complicated than the old one, the calculations usually require rather more complicated mathematics.

This new generalized law of motion is put forward as the basic law of special relativistic dynamics which enables the motion of any body to be found even if the body is moving at a speed near that of light. It is clearly asymptotic to the basic classical law for low speeds but so far we have given no reason why Newton's second law should be generalized in this particular way. This is because no reason can be given except that the predictions of the law are experimentally correct. Therefore we shall now consider the experimental evidence for the new generalized law.

Experimental evidence for the new basic law of dynamics must obviously be sought in situations where the speeds involved are very high. The speeds required to show significant differences between classical dynamics and the dynamics of special relativity must exceed 0·01 Ls/s (or 3×10^6 m/s) and, as can be seen from Fig. 2.1, such speeds exceed by a factor of about a thousand the maximum speeds which can be attained even with the most powerful modern space rockets. Also the speed of the earth in its orbit relative to the sun is only about 3×10^4 m/s, so that there are no large bodies known which can be observed moving at sufficiently high speeds for the new laws to be tested. The only bodies which can be observed moving at speeds of the required order of magnitude are very small atomic particles like the electrons and pions which were studied in some of the experiments described earlier in this book.

Sometimes nature itself is kind enough to provide such particles moving at speeds comparable with that of light. The cosmic rays,

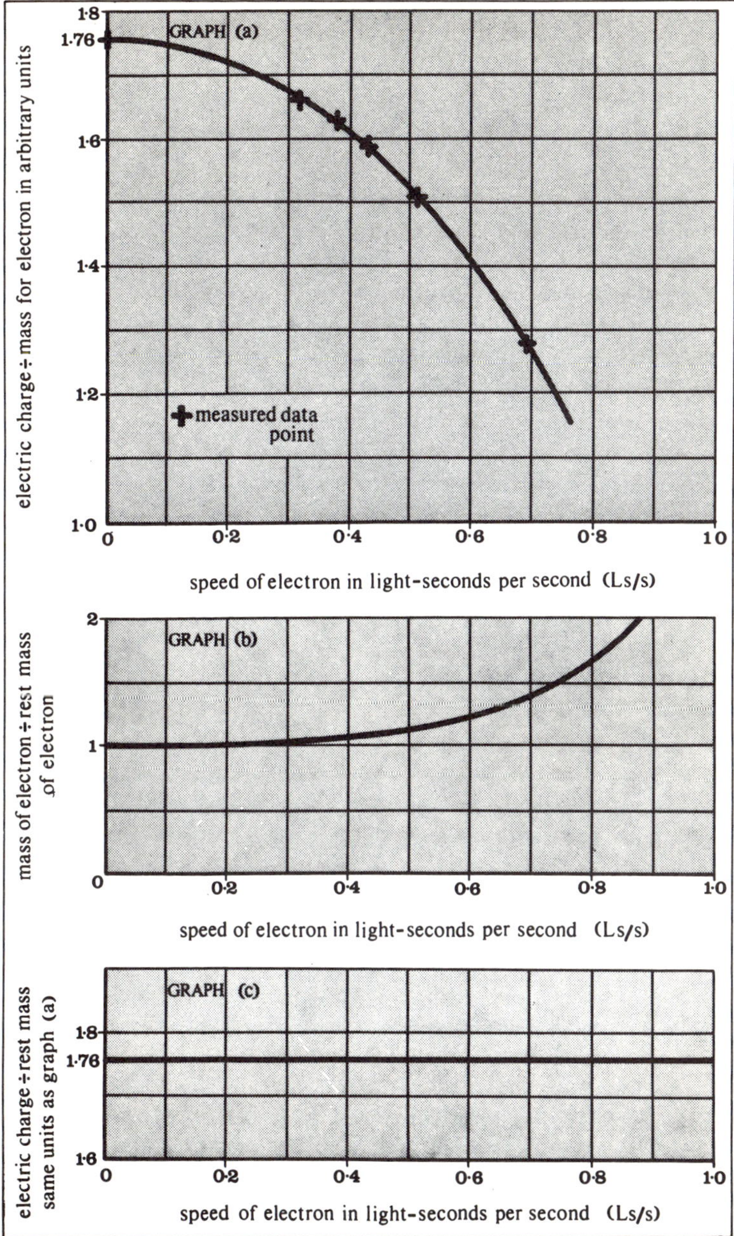

Fig. 5.2. Graphs showing (a) the ratio of charge to mass for an electron, (b) the ratio of mass to rest mass predicted by special relativity and (c) the ratio of charge to rest mass for an electron, all as a function of speed. Graph (c) is obtained by point-by-point multiplication of graphs (a) and (b).

which continuously bombard the earth from outer space, contain many such fast particles, while some of the particles emitted by radioactive atoms are moving fast enough to show relativistic effects; in fact, the electrons used in the experiment described earlier in this chapter (Fig. 5.2) were obtained from a radioactive source. However, by far the most extensive amount of experimental work with fast particles has been carried out in connection with the design and operation of man-made particle accelerators like the one at CERN which has already been mentioned once or twice. Such accelerators have been built in many laboratories all over the world in order to advance our knowledge of the detailed structure of the atoms which make up all matter, and in particular to study the forces which are observed when particles of matter are very close together (such forces are called nuclear forces and are another kind of force which can be included in the list of types of force given in Chapter 3). The principal way in which this is done is to see what happens when a fast-moving particle collides with an atom and, naturally enough, the faster the particle is moving, the more diverse are the things that happen in these collisions and the more interesting are the conclusions which can be drawn about the structures of the atoms involved. Consequently there has been a natural tendency for scientists to design the accelerators so that they provide particles which are moving as fast as possible.

Nearly all accelerators work by exerting combinations of electric and magnetic forces on electrically charged atomic particles. The charged particles are produced in some kind of source and are then subjected to a series of electric and magnetic forces of such directions and magnitudes that they follow closely defined paths in space at continually increasing speeds. They finally emerge from the accelerator travelling at very high speeds and are directed on to a target so that the collisions mentioned above can be observed and studied.

These accelerators are not easy to build and the designers must make many careful and detailed calculations of the forces to be applied and the consequent paths followed by the particles before a successful design is obtained. In every case it is found that the motions predicted by the basic law of motion of special relativity, rather than those predicted by Newton's classical law, are the motions which are actually found to occur in the machines whenever the speeds of the particles exceed about 3×10^6 m/s. As an example, consider the cyclotron, which was one of the first accelerators to be built. In a cyclotron the particles start at the centre of a circular apparatus and move in a spiral path under the influence of suitable electric and magnetic forces until they emerge at high speeds from an aperture at the circumference of the apparatus. The early cyclotrons were designed to give particles which travelled at speeds appreciably less than 3×10^6 m/s and so classical dynamics was adequate for their design. However, when scientists attempted to increase the speeds of the particles from a cyclotron up to and beyond this value merely by increasing its size (which should have been successful according to classical dynamics), then it was found that the machine did not work. The particles collided with the walls of the apparatus instead of emerging in the calculated place, and it was only when the new special relativistic dynamics was used that successful machines were constructed.

There are now probably more than a hundred accelerators of various types in operation all over the world, all designed using relativistic dynamics, and many of them have been in almost continuous successful operation for a number of years. None of these machines could function successfully if the classical equations were correct. Therefore the weight of experimental evidence in favour of the special relativistic dynamics is overwhelming. Also many different types of particle, as well as electrons, have been accelerated in these machines so that the new law is not just applicable to electrons, as might have been suspected from the experiments described earlier in this chapter.

The successful operation of all these accelerators also provides additional evidence for the limiting speed of particles being the same as that of light and for the assumption that the mass of a particle varies with speed but its electric charge remains constant (made in the experiment on mass variation described earlier in this chapter), since both these points are relied on in accelerator design.

Force has thus been redefined in a way that is consistent with special relativity and has been shown to be a useful quantity in relativistic dynamics. Now let us consider some other useful classical quantities which are defined in terms of force such as work and energy, and in particular kinetic energy. These quantities can still be defined in terms of force in exactly the same way as they were defined in classical physics (see Chapter 3), provided that force is defined in the new way discussed in this chapter. Work can still be defined as the product of force and distance moved in the direction of the force, energy as the amount of work done on a body or capable of being done by a body, and kinetic energy as energy of motion. All these quantities can be shown to be useful in relativistic dynamics and to show the required asymptotic behaviour with respect to the classically defined quantities when the speeds involved are low enough. It is easier to discuss these quantities and to show the similarities and differences between the classical and special relativistic definitions if a simple example is considered in detail. It will also turn out that the example will suggest a totally new special relativistic effect which has no parallel in classical physics.

Consider a body which is initially at rest and is acted on by a constant force for an indefinite period of time. Fig. 5.3 and 5.4 show how the momentum, speed, mass, position and kinetic energy of the body vary with time both according to the laws of classical dynamics and according to the new special relativistic laws of dynamics; the dashed lines in these figures show the classical predictions while the solid lines show the special relativistic ones. Many of the classical predictions have already been described in Chapter 3 and reference will be made to the appropriate figures as Fig. 5.3 and 5.4 are discussed in detail.

Fig. 5.3(a) shows how the momentum of a body varies with time if the constant force is 1 N (compare line F_1 in Fig. 3.5); here the classical and relativistic lines coincide at all times and the lines will be the same for any value of the rest mass of the body.

Fig. 5.3(b) shows how the speed of the body in light-seconds per second varies with time if the rest mass of the body is $3{\cdot}33 \times 10^{-9}$ kg (compare Fig. 3.2). This particular rest mass has been chosen as then the classical speed reaches 1 Ls/s after 1 s. This means that any differences between classical and relativistic dynamics are clearly shown on the diagram; however, there is no special significance in this choice of rest mass and similar diagrams for any other mass can easily be drawn. The relativistic curve can be calculated fairly easily from the generalized relativistic form of Newton's second law of motion and the relativistic relation between mass and speed shown in Fig. 5.2(b); it is clearly asymptotic to the classical prediction at low speeds and approaches the relativistic limiting speed of 1 Ls/s as time increases.

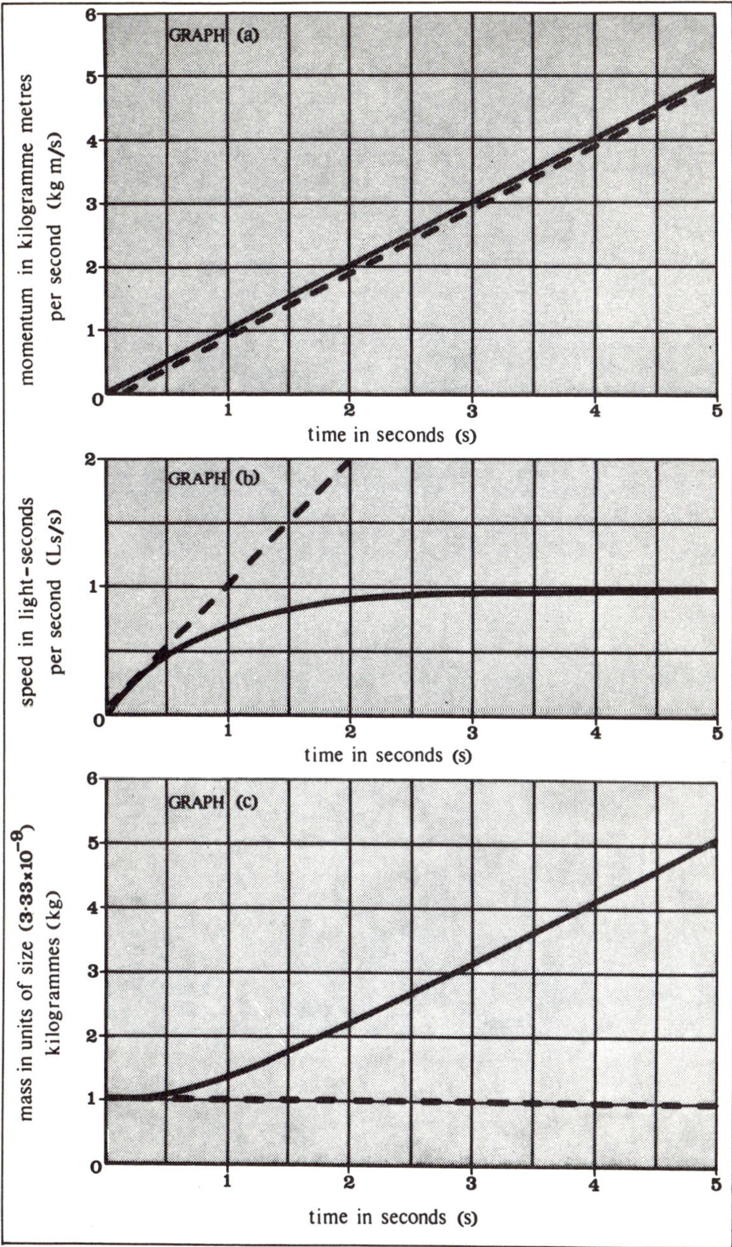

Fig. 5.3. Graphs showing the variation with time of (a) the momentum, (b) the speed and (c) the mass of a body of rest mass $3{\cdot}33 \times 10^{-9}$ kg, initially at rest, which is subjected to a force of 1 N. The dashed lines and the solid lines show respectively the predictions of classical and special relativistic dynamics.

Fig. 5.3(c) shows how the mass of the body varies with time; the relativistic curve can be obtained directly from Fig. 5.3(a) and 5.3(b) since we know that the momentum values in (a) can be obtained by point-by-point multiplication of the speed curve in (b) and the mass curve in (c) (for example, after 1 s the momentum is 1 kg m/s from (a), the speed is 1·41 Ls/s from (b), and therefore the mass is $(1 \div 1{\cdot}41)$ or 0·71 units as shown in (c)). Since the speeds in Fig. 5.3(b) are expressed in light-seconds per second rather than metres per second, the masses in Fig. 5.3(c) will not be in kilogrammes but in a rather unusual unit $3{\cdot}33 \times 10^{-9}$ kg in size; thus the body being considered has a rest mass of 1 of these unusual units.

Fig. 5.4(a) shows how the position of, or distance moved by, the body varies with time (compare Fig. 3.7(a)); the distances are expressed in light-seconds. The curves in Fig. 5.4(a) are obtained from the curves in Fig. 5.3(b) in exactly the same way as Fig. 3.7(a) was obtained from the solid line F_1 in Fig. 3.3; that is, the slope of either curve in Fig. 5.4(a) at any time is always equal to the speed of the body at that time as given by the equivalent curve in Fig. 5.3(b) (for example, the relativistic curve in Fig. 5.4(a) becomes a straight line of slope 1 Ls/s when the equivalent curve in Fig. 5.3(b) approaches the constant speed of 1 Ls/s). Fig. 5.4(b) shows how the applied force varies with time (compare Fig. 3.7(b)). Finally, Fig. 5.4(c) shows how the kinetic energy acquired by the body varies with time (compare Fig. 3.7(c)); it is obtained by point-by-point multiplication of the curves in Fig. 5.4(a) and 5.4(b) in just the same way as Fig. 3.7(c) was obtained from Fig. 3.7(a) and 3.7(b). Since, however, the distances in Fig. 5.4(a) are in light-seconds rather than in metres, the kinetic energies in Fig. 5.4(c) will not be expressed in joules but in another rather unusual unit 3×10^8 J in size.

The six graphs of Fig. 5.3 and 5.4 show quite clearly some of the similarities and differences between classical and special relativistic dynamics. They also show that where the predictions of the two theories differ, they are asymptotic to one another for low speeds. There is also, however, something much more interesting and suggestive about these graphs as we can see if we look carefully at the relativistic curves in Fig. 5.3(c) and Fig. 5.4(c). If we do this we can see that these curves are identical, except that the mass curve starts at 1 unit (the rest mass) at zero time whereas the kinetic energy curve starts at zero at zero time. Another way of putting this is to say that

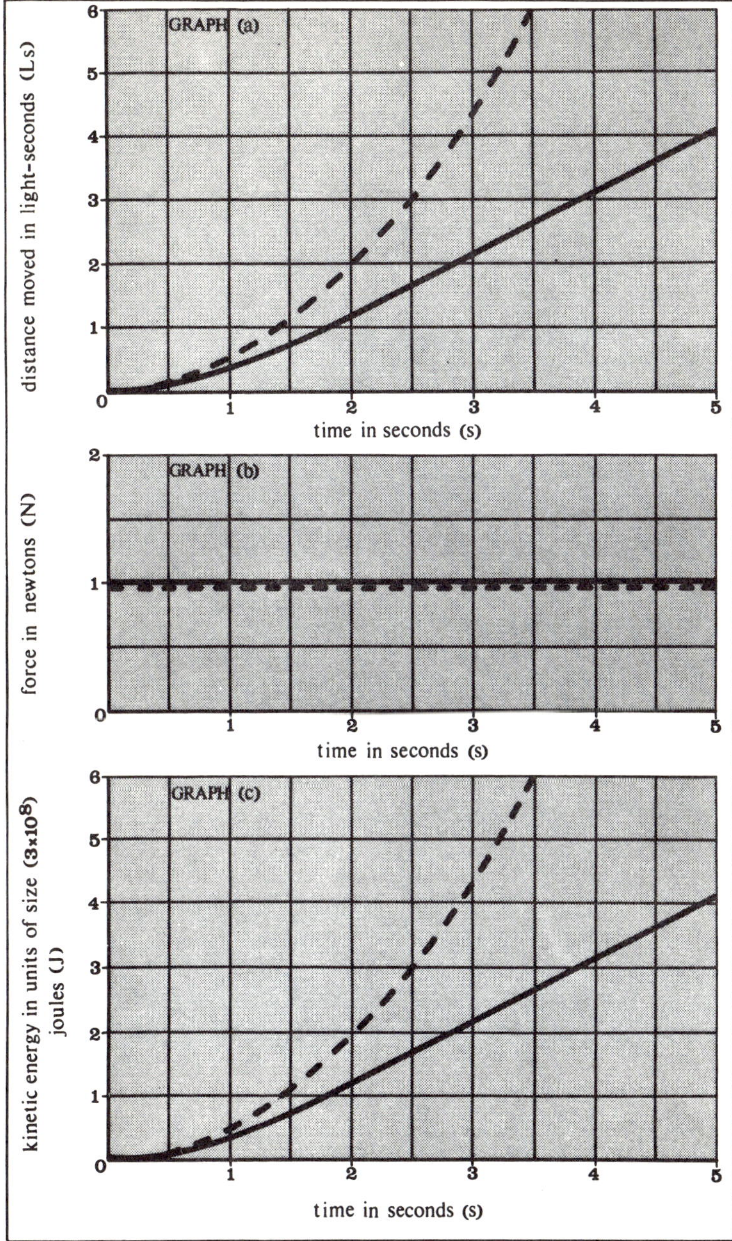

Fig. 5.4. (a) Distance moved as a function of time for the body of Fig. 5.3. (b) Force acting on the body as a function of time. (c) Kinetic energy of the body as a function of time. The dashed lines and the solid lines show respectively the predictions of classical and special relativistic dynamics. The graphs in (c) are obtained by the point-by-point multiplication of the graphs in (a) and (b).

the increase in mass of the body at any time (or the difference between the mass of the body at that time and its rest mass) is exactly equal to the kinetic energy of the body at that time.

Now we have only considered one example of the application of relativistic dynamics and this example is something of a special case since we have used a unit force (1 N) and, in our unusual units, a unit mass (1 unit of $3{\cdot}33 \times 10^{-9}$ kg size). It is not difficult, however, to draw curves similar to those of Fig. 5.3 and 5.4 for any values of force or mass. And when we do this, we always get the same result—the kinetic energy at any time is always exactly equal to the relativistic increase in the mass (that is, to the difference between the mass at that time and the rest mass). This result is always true, provided that we measure kinetic energy and mass in the unusual units which we have defined above or in any other consistent set of units in which the speed of light has the value 1 unit.

If units are used in which the speed of light is not 1 unit, then conversion factors must be introduced, but this is a point of convention rather than a point of principle. It does not affect the surprising conclusion that the kinetic energy or energy of motion of a moving body is exactly equivalent to the extra mass acquired by that moving body. This conclusion is valid even at low speeds, but since 1 J of kinetic energy (or $3{\cdot}33 \times 10^{-9}$ of our unusual units of energy) is equivalent to a mass increase of only $1{\cdot}11 \times 10^{-17}$ kg (or $3{\cdot}33 \times 10^{-9}$ of our unusual units of mass), it is very difficult to detect this change in experiments involving ordinary everyday masses and amounts of energy.

The equivalence between kinetic energy and extra mass was surprising enough since mass and energy had previously been regarded as separate distinct physical quantities. However, the next step or guess taken by Einstein was even more surprising. He argued that since the extra mass acquired by a moving body was found to be equivalent to a form of energy, one could consider the rest mass of a stationary body also to be a form of energy. The total energy of the body at any time would then be exactly equal to its mass in our unusual units. Support came for this suggestion when detailed calculations, using the Lorentz transformation, showed that total energy, when defined in this way, was conserved in relativistic dynamics in the same way as energy was conserved in classical dynamics; different inertial observers are found to agree on the conservation of total energy in any series of events viewed by all of them, even if they disagree on the amounts of total energy involved.

However, the equivalence between mass and energy suggested by Einstein has been found to be much more complete than the last paragraph implies. A true equivalence between rest mass and other forms of energy would require that rest mass could under some suitable circumstances be converted into the equivalent quantity of some other form of energy and also that the reverse process, the conversion of some other form or forms of energy into the equivalent amount of rest mass, should also occur. Then rest mass would be seen to behave just like any other form of energy. Since these ideas were put forward, a large number of experiments have shown that this is just what does happen under suitable circumstances. We shall now consider some of these experiments in more detail.

Most of these experiments have been the sort of collision experiment which was briefly mentioned earlier in connection with particle accelerators. In a typical collision experiment a fast-moving particle collides with a stationary particle with such violence that both particles break up and cease to exist in their original forms. Very often, the aftermath of such a collision is that two or more different particles are formed and then proceed to fly apart from the point at which the collision took place. The particles involved in these collisions are usually the very small particles which constitute the nuclei of atoms and the effect of such a collision is that nuclei of one or more kind of atom change into nuclei of some other kinds of atom. The original nuclei are said to react with one another during the collision, and the whole of such a collision process is called a nuclear reaction.

Careful measurements are usually made of the kinetic energy of the incident nucleus, the kinetic energies of the two or more product nuclei, and the amount of any other form of energy which is produced during the collision. When this is done it is almost always found that energy is not conserved in the collision; the energy carried away by the products of the collision is almost always either less than or greater than the initial energy of the incident nucleus. The principle of conservation of energy is not applicable to these nuclear reactions. However, the rest masses of most nuclei have been measured to an extremely high degree of accuracy using instruments called mass spectrometers and it is found that rest mass is not conserved in such reactions either. If excess energy is produced in a reaction, then some rest mass is found to disappear in that reaction, while if energy disappears in a reaction then the reaction products are found to have more rest mass than the original nuclei. Furthermore, the amount of rest mass which appears or disappears in these reactions is always equal, within the limits of accuracy of the measurements, to the amount of energy which has been lost or gained (provided that the rest masses and energies are expressed in the unusual units which we have been using in this chapter).

Many thousands of nuclear reactions have been studied in detail in this way and it is always found that the total amount of rest mass and energy together is the same before and after the reaction. In some cases rest mass disappears and is converted into energy, while in others energy disappears and is converted into rest mass, but always in exactly equivalent amounts.

The early experiments on nuclear reactions were on a fairly small scale and the total amounts of energy released were usually very small too. More recently, however, the conversion of rest mass into energy has been carried out in nuclear weapons and nuclear power stations on a scale which is so large that it has posed many extremely difficult problems in both the technical and political fields. Indeed it may yet turn out that these problems, of both types, are too difficult for us to handle.

Similar nuclear reactions on an even larger scale are now known to be one of the main sources of the energy radiated by the sun and other stars. Our sun, for example, is converting rest mass into energy at the rate of more than 10^9 kg every second, which, however, is a negligible loss of mass compared with its total rest mass of over 10^{30} kg. These large-scale phenomena are dramatic examples of the conversion of rest mass into energy.

The reverse process has also been most convincingly demonstrated by experiments in which a fast particle is rapidly decelerated. The kinetic energy lost by the particle during the deceleration is sometimes transformed into the rest mass of one or more new particles. In other words, new particles are sometimes created from the kinetic energy lost by the original fast particle. These new particles are observed, and their properties measured, by the well-established methods used to detect any other kind of atomic or nuclear particle. Particles which have been created in this way include electrons, pions and nuclei of some of the lighter atoms, as well as some kinds of particle which were previously unknown. Obviously if the kinetic energy of the original particle is increased then such experiments are likely to lead to the production of a wider range of particles since more initial kinetic energy means that more 'potential rest mass' is available for particle creation. The success of this type of experiment is one reason for the construction of ever larger accelerators capable of producing particles with ever higher kinetic energies.

All these experiments have amply confirmed the equivalence and interconvertibility of rest mass and energy, and so much so that the classical conservation of energy principle can now be generalized into the special relativistic principle of the conservation of total energy with rest mass included as one of the possible forms of energy.

Other measurements which are quite often made when nuclear reactions are studied are the directions in which the products of the reaction move. In this way the momentum of each particle involved can be measured, and it is found that momentum is always conserved in such reactions, provided that the relativistic definition of momentum is used. In one or two cases, the apparent failure of the principle of conservation of relativistic momentum has led to more careful experiments which have shown that the 'lost' momentum has in fact been carried off by a hitherto unknown particle.

Other parts of special relativity theory which are supported by the detailed study of nuclear reactions are the Lorentz transformation and the related transformations for velocity, acceleration and many other quantities. These transformations are constantly used in the analysis of the results of nuclear reactions. Therefore the satisfactory and consistent way in which such reactions can now be analysed provides further evidence in favour of the relativistic transformations.

Thus we have seen that the conservation principles, when suitably modified, are retained in special relativity as being invariant for all inertial observers, despite the fact that the total energy or the total relativistic momentum of any body is not invariant under a Lorentz transformation. However, there is another important quantity which is found to be invariant under a Lorentz transformation, and the magnitude of this quantity is calculated in a way similar to that used to calculate the square of the other invariant of special relativity which we have already mentioned, the interval between two events. If we calculate the square of the total energy of a body and subtract from it the square of the momentum of the body, then we find that the resulting quantity is invariant under a Lorentz transformation, even though the total energy or momentum is not. If the units are chosen so that the speed of light has the value 1 unit, this new invariant is found to be equal to the square of the rest mass of the body. So the rest mass of a body which had perhaps been a little devalued by special relativity when it was shown to be just another form of energy, now has regained some of its importance, since it is one of the few attributes of a body which appears the same to all inertial observers.

The theory of special relativity is thus a theory which has a vast amount of experimental evidence to support it. It is asymptotic to the classical relativity theory for speeds small compared with the speed of light; therefore all the experimental evidence which supports classical relativity also supports special relativity. We have also seen that its predictions in regions of experience beyond the realm of common-sense are confirmed by a wide variety of experimental evidence. The correctness of some of its predictions is so well attested that the theory has long ceased to be of interest only in the interpretation of the esoteric experiments of some research scientists. It is now crucial to the success of some large-scale engineering projects such as the design and construction of nuclear power plants, nuclear weapons and particle accelerators. The theory has been successfully applied to situations involving electromagnetic forces, nuclear forces and gravitational forces, which comprise the three main types of force that are known to exist in nature (all types of mechanical force can be shown to be electromagnetic in origin when they are studied in terms of the detailed interactions between atoms). The theory of special relativity therefore has a much wider applicability than Einstein first thought and is as well supported by experimental evidence as almost any part of physics.

6
How Special Relativity Was Really Born

Special relativity has been presented in Chapter 4 as if its chief inspiration was a desire to look more closely and rigorously into the nature of space and time. This is how the theory is usually presented and in many ways it is a misleading account of the origin of the theory. In this chapter we shall try to indicate the background of ideas in physics which were current in the early years of this century and which form the scientific environment in which special relativity emerged. Hopefully this will provide a brief interlude sandwiched between the rigours of the special theory and the general theory.

One of the major scientific achievements of the nineteenth century had been the progress in our understanding of electricity and magnetism. Many scientists had contributed to this understanding which had been reached by the sort of combination of theoretical and experimental work discussed in Chapter 2. This progress culminated around 1860 in the electromagnetic theory of Maxwell. Maxwell brought together all the preceding work on electricity and magnetism into a systematic mathematical theory and by manipulating his equations he was able to show that certain kinds of electrical experiments should give rise to the transmission of electrical and magnetic energy through the surrounding space in a manner which resembled the transmission of other forms of energy by means of water waves or sound waves. Maxwell gave the name electromagnetic waves to these hypothetical waves and he was able to deduce the speed at which they should move in terms of electrical and magnetic constants which had already been measured. This calculated speed was exactly equal to the speed of light which suggests that light, already considered to be some kind of wave, is in fact a form of electromagnetic wave. The theory was crowned in 1887 by the detection of electromagnetic waves in an experiment involving an electrical circuit. These waves also travelled at the calculated speed and had all the other properties predicted by Maxwell; they are, of course, the radio waves which are so familiar today. Thus Maxwell's theory was triumphantly vindicated.

Soon, however, it was realized that this great success of nineteenth-century physics was not quite consistent with that earlier great success of classical physics, Newton's dynamics. In particular some peculiar results appeared when the Galilean coordinate transformation of classical relativity was applied to Maxwell's theory. His equations were found not to be invariant under a Galilean transformation and the application of this transformation predicted effects which were not observed, such as different speeds for electromagnetic waves travelling in different directions with respect to a moving observer. Around 1900 Lorentz was studying this problem and by a process of trial and error he arrived at a coordinate transformation for which Maxwell's equations were invariant; this transformation is the familiar Lorentz transformation mentioned in Chapter 4. Lorentz himself, however, gave no reason why it should be true except that its application to Maxwell's equations left them invariant.

Einstein was not aware of this work of Lorentz when he wrote his first paper on special relativity in 1905. However, it is clear from the introduction to his paper and from his later writings that it was the inconsistencies between electromagnetism and classical relativity that had first led him towards his revision of the accepted ideas about space and time. In fact he derived quite independently the same coordinate transformation as Lorentz in the way we have described in Chapter 4. He also showed in that same paper that Maxwell's equations were invariant under this new transformation. There is no reason why Einstein's ideas on space and time should not have been developed quite independently of anything in electromagnetic theory but this was not in fact what occurred. The stimulus for special relativity was provided by the general ferment of ideas at that time and it seems likely that the theory or something very like it would have soon emerged even if Einstein had not lived. This does not detract from Einstein's achievement but perhaps indicates something about the way in which new scientific concepts arise and how the emergence of such concepts often depends very much on the scientific environment created by the achievements of a large number of earlier workers.

Another interesting connection exists between electromagnetism and special relativity and this was also briefly mentioned in Einstein's original paper. It can be shown that all the laws of electromagnetism, which were established with so much labour over so many years, can be derived from a few simple assumptions together with the principles of special relativity. Starting from the well-known law for the force between two electric charges (Coulomb's law) and assuming that electric charge is invariant under a Lorentz transformation (see Chapter 5), it is possible to deduce all the laws of electromagnetism including the very existence of magnetic forces. Thus on this basis one could say that magnetism itself is a relativistic effect.

With the wisdom of hindsight we can perhaps not be too surprised at this close connection between electromagnetism and special relativity. We know from Maxwell's work that the laws of electromagnetism imply the existence of waves travelling at what we now know to be the limiting speed of special relativity. Thus experiments involving ordinary electric and magnetic effects, which can be performed with the simplest of apparatus, necessarily involve a kind of motion which is so fast that we might expect classical ideas to be inapplicable to it. It is true that this motion is not obvious and is not the motion of a body with a definite rest mass, but nevertheless it does result in the transfer of energy from one place to another and so special relativity ought to be relevant. And, as we have seen, special relativity is implicit in the equations of electromagnetism.

Thus the existence of magnetic forces, the truth of the well-tested laws of electromagnetism and the many types of electromagnetic wave that have now been observed can all be taken as indirect evidence for the special theory of relativity. So this chapter, which was intended as an interlude from the main theme of the book, has ended by adding to the already considerable experimental support for special relativity.

7

General Relativity, Space-Time and Gravitation

The general theory of relativity was first published by Einstein in 1916, more than ten years after his main work on the special relativity theory was completed. This long delay in the development of the general theory at a time when Einstein was at the height of his powers gives some idea of its much greater difficulty compared with the special theory. It is not at all a straightforward generalization of the special theory but is really almost a totally new theory and it is an unfortunate accident that the two theories have such similar names and are often spoken of in the same breath. The general theory does borrow and generalize some ideas from the special theory but it starts from very different principles and arguably differs more from special relativity than special relativity does from classical relativity. It is much more ambitious, much less fully worked out and has fewer points of contact with experiment than the special theory. Consequently the theory is rather abstract and mathematical and it is not easy to present it in a non-mathematical way. However, an attempt will be made in this chapter to present the main ideas of Einstein's general theory, to discuss the experimental evidence relating to it and to indicate the relationship between the special and the general theory, leaving for the next chapter a more general discussion of some recent ideas concerning the present status of general relativity.

The general theory of relativity seeks to derive a theory of the physical world which can employ coordinates measured by any kind of observer, not just inertial observers, and which also incorporates gravitation in a fundamental way totally unlike the way it has hitherto been treated in theoretical physics. The first of these objectives would be task enough for any new theory, but in Einstein's approach it is really secondary compared with the challenge of incorporating gravitation into his theory. The reasons for selecting these two objectives, and in particular the second one, are not self-evident and we will now discuss why they were chosen to be the main aims of the theory.

In order to start the discussion let us return to classical relativity and see how it deals with the problem of the non-inertial observer. This may appear to be something of a digression from the main theme of this chapter but in fact it brings us fairly quickly to the crucial considerations which were in Einstein's mind when he was developing the general theory. As an example of a non-inertial observer, consider an observer who is falling freely near the surface of the earth. Let us assume that he is falling through an enormous vertical tube from which all the air has been removed so that any forces due to air resistance can be neglected. Such an observer has an acceleration with respect to the earth's surface which is produced by the gravitational attraction of the earth on him. This acceleration is practically constant at any point in the vicinity of the earth's surface provided that the distance of the point from the earth's surface is small compared with the radius of the earth. If this observer carries another object with him and at some point in his motion lets go of the object, then he observes that the object remains at rest with respect to him.

An inertial observer on the earth's surface would interpret these events by saying that the non-inertial observer and the object were falling with exactly the same acceleration. This is a well-known observation first made by Galileo and since generalized by subsequent experiments in the statement that any body placed at a given point in space will experience the same acceleration with respect to an inertial observer if the only forces acting on it are gravitational ones. The inertial observer explains the motion of the object by saying that the object experiences a gravitational force which is directly proportional to its mass and hence, by Newton's second law of motion, it experiences an acceleration. The non-inertial or freely falling classical observer will still say that the object is experiencing the same gravitational force but according to his measurements the object is at rest. Hence as far as he is concerned Newton's second law of motion is not applicable, since the object experiences a finite force but has no acceleration.

In order to retain Newton's second law of motion for such a non-inertial observer, classical physics introduced an apparent or fictitious force of such a magnitude and direction that the law would still seem to be true to him. Such a force in this case must obviously be equal and opposite to the gravitational force so as to cancel it exactly and leave the object at rest. In other words the fictitious force assumed to be acting on the object must be equal in magnitude to the mass of the object multiplied by the acceleration of the non-inertial observer and directed in the opposite direction to that acceleration. Such fictitious forces are called *inertial forces* and they can be introduced in order to save Newton's second law for any non-inertial observer. Other examples of inertial forces are centrifugal forces, which are involved in describing the motion of uniformly rotating bodies, and Coriolis forces, which are used in calculating the paths of missiles or shells fired from one point on the earth's surface to another or in explaining the directions of the trade winds in the earth's atmosphere.

Inertial forces are considered by the non-inertial observer to act on all bodies and they are always equal to the mass of the body multiplied by the acceleration of the observer. These inertial forces may seem a somewhat artificial way of preserving the laws of physics intact for all observers, but their use has been shown to give consistent results which agree with the experimental observations in all cases for which the speeds involved are low enough for classical concepts to be applicable. Also their properties were one of the considerations in Einstein's mind when general relativity was being conceived.

Another such consideration was the unique nature of gravitational forces compared with all other forces. In both classical and special relativity gravitational forces are treated in much the same way as any other force but, as Einstein realized, gravitational forces have some properties which other types of force do not possess. Suppose we attempt to verify Newton's first law of motion by eliminating all the forces acting on a body and then measuring its motion to see if it does indeed then move with constant velocity. We can remove all mechanical forces by making sure the body makes no direct contact with any other body. We can remove all electrical or magnetic forces by making sure that the body itself is an electrical insulator and carries no electric charge or permanent magnetism. But nothing we do to the body can eliminate the gravitational force acting on it. The gravitational force cannot be removed by any local measures but only by the extremely drastic step of removing all other bodies to enormous distances from the body in question.

This is one way in which gravitational forces differ from all other forces but it is not the only unique characteristic of gravitation. The gravitational force exerted on a body placed at any point in space is directly proportional to the mass of the body (see Chapter 3) but depends on no other property of the body at all. Also the gravitational force exerted on each of two bodies placed successively at the same point in space will be the same if they have the same mass, no matter how different the materials of which the bodies are made. Further than that, the same acceleration is produced by the action of gravitational forces on any body placed at a particular point in space; the acceleration depends neither on the mass of the body nor on the material of which it is made. Suppose a mass of 1 kg were to experience a gravitational force of 1 N at a particular point in space; this would give it an acceleration of 1 m/s (see Chapter 3). A mass of 2 kg at the same point would experience a gravitational force of 2 N which would still give it an acceleration of 1 m/s. Similarly a 3 kg mass would experience 3N, a 4 kg mass 4N, and so on, thus giving the same acceleration, 1 m/s, to any body placed at this point in space.

It is important for general relativity to know if this equality of gravitational acceleration for different masses is exact or just approximate. Recent extremely precise experiments, which are the modern equivalent of Galileo's observations, have shown that the gravitational acceleration experienced by any body at a given point in space is the same within an accuracy of at least one part in one hundred thousand million (1 part in 10^{11}), which is an extremely high accuracy by any standards. In other words, gravitation appears to be a phenomenon which is characteristic of the locality or region in space in which a body finds itself rather than a characteristic of the body itself. And we shall see that this is just the property that Einstein's general relativity attributes to gravitation.

We can now take the next step towards the central idea of general relativity if we consider a striking similarity between gravitational forces and the inertial forces introduced earlier in this chapter. This is that they are both forces which are completely independent of the nature of the body involved but which depend only on its mass. This similarity led Einstein to suggest that gravitational and inertial forces were indistinguishable from one another in their effects on any measurements made inside a closed vehicle by an observer who had no contact with anything outside the vehicle. This suggestion is called the principle of the equivalence of inertial and gravitational forces, or sometimes simply the equivalence of inertia and gravitation, and is one of the basic principles on which general relativity depends. The principle is probably best illustrated by a series of simple examples.

Consider an observer inside a closed vehicle which is stationary on the earth's surface. Suppose that all the air has been removed from the vehicle and that the observer is provided with suitable clothing and a life-support system so that he can carry out measurements and experiments. Let him carry out the simple experiment (experiment 1) of releasing an object and observing its subsequent motion. He will observe that the object moves with a constant acceleration attributable to the gravitational attraction of the earth. This acceleration is called the acceleration due to the earth's gravity at the particular point on the earth's surface and is usually about 9·8 m/s^2; let us suppose in our example that it is exactly 9·8 m/s^2. Now imagine that the closed vehicle with the observer inside it is put into an enormous vertical evacuated tube, like the one mentioned earlier in this chapter, and allowed to fall freely. If the observer now attempts to carry out the same simple experiment of releasing an object and observing its subsequent motion (experiment 2), he will observe that the object remains at rest. Now imagine that the vehicle and observer are transported to a region in space which is far from all other bodies that gravitational forces can be neglected. If the observer again attempts to carry out the same experiment (experiment 3), he will still observe that the object remains at rest. Finally, imagine that the vehicle, while still in this remote region of space, is given an acceleration of exactly 9·8 m/s^2 by means of suitable external rocket engines and that the observer then repeats the same experiment again (experiment 4); he will now observe that the object moves with an acceleration of exactly 9·8 m/s^2.

Consider first experiments 1 and 4. In experiment 1 we, from our viewpoint outside the vehicle, would say that the observed acceleration was due to a gravitational force while in experiment 4 we would say that it was due to an inertial force. But to the observer inside the vehicle the two situations are identical. He is unable to distinguish between what we call a gravitational force and an inertial force by this or any other experiment carried out solely inside his vehicle. This is one very simple example of the application of Einstein's principle of equivalence.

Now consider experiments 2 and 3. In experiment 2 we explain the absence of motion of the object by saying that the gravitational and inertial forces are equal and opposite and thus cancel each other, while in experiment 3 we would say that no forces of any kind were acting and that therefore the object remains at rest. Once again, however, the observer inside the vehicle cannot distinguish between these two situations since he cannot separate the inertial and gravitational forces and measure them one at a time. To him the two situations are equivalent and this is another simple example of the principle of equivalence. The sort of situation arising in experiments 2 and 3 will also arise whenever the vehicle is allowed to move freely with its engines switched off, no matter where it may be. The inertial and gravitational forces will always exactly balance each other and the observer will not be able to tell if he is in a region in which gravitational effects exist or not. Such a motion is often called 'free fall' and is now a common experience in manned space flights; the effect we have been describing is usually called 'weightlessness'. Similarly the situation in experiment 4 is exactly like that which occurs in manned space flights when the rocket engines are switched on. The observed phenomena are exactly like those we have described in our simple example.

The principle of equivalence therefore indicates that inertial and gravitational forces are very much the same sort of thing and suggests that a satisfactory physical theory ought to treat them similarly rather than to distinguish between them. Another connection between inertia and gravitation is suggested by the attempted verification of Newton's first law of motion, or the principle of inertia, which we described earlier in this chapter while discussing the unique nature of gravitational forces. Then we found that we could not directly test Newton's first law, since we could never remove the gravitational force and so observe pure inertial motion at constant velocity. This situation is rather unsatisfactory since Newton's first law is our basic definition of the natural motion of a body and it is surely desirable that something as fundamental as that ought to be capable of being tested.

These considerations together with all the other factors which we have discussed in this chapter led Einstein to take a step that was drastic even for him. He proposed that the definition of the natural unforced motion of a body should be revised. Instead of regarding this natural motion as motion with constant velocity, or inertial motion, he proposed that the natural unforced motion of a body should be defined as being the motion of a body under free fall conditions, that is under inertia and gravitation combined. This was an audacious generalization of the basic law of motion which had been at the very centre of dynamics for so long and had been retained in his own successful special relativity (see Chapter 5). And it was a generalization which, once it was made, could be seen to have considerable advantages. It retains the idea of a natural motion for a body and also retains the concept of a force as something which causes a body to deviate from this natural, free fall motion. But it abolishes the idea of a gravitational force since the free fall motion of the body due to gravitation and inertia is now its natural motion. Thus gravitation and inertia are combined, gravitation is indeed something special and not like other forces, and the difficulties about verifying Newton's first law also disappear. One disadvantage is that the natural motions which are possible under what we might call Einstein's first law of motion are far more varied and complicated than the constant velocity, straight-line natural motions of Newton's law. And it is one thing to state this new law but quite another to work out a detailed theory which would enable the actual motions of real bodies to be calculated.

The new detailed theory must also be asymptotic both to the classical theory of gravitation and to special relativity, since both these theories have had considerable success in accounting for large amounts of experimental data. Thus the task facing Einstein was very difficult and it was a great achievement when he was able to produce such a theory. As a kind of bonus, it turned out that the theory he produced to unite gravitation and inertia almost involuntarily or automatically satisfied his other aim, the abolition of the restriction to coordinates measured by a particular class of observers. We shall now try to give some account of the main ideas of the theory as presented by Einstein.

Einstein's general theory retains some of the ideas and terminology of special relativity. In particular it retains the idea of a space-time framework for an observer's view of the world in which events occur with positions in space-time that can be characterized by four co-ordinates. Secondly the life or career of any body or observer can still be represented in space-time by a world-line made up of the sequence of all events experienced by the body or observer. And thirdly, it is still possible to define the interval between two events in a manner similar to that used in special relativity, provided that the events are very close together in space-time. The initial question that Einstein set himself was to try and find out what particular world-line or kind of world-line would correspond to the world-line of a freely or naturally moving body, using of course the new definition of a naturally moving body which we have called Einstein's first law of motion. And in order to see how he answered that question we shall need to make what may seem to be a lengthy digression.

Let us first go back to the space-time of special relativity and consider Fig. 7.1, which shows a space-time diagram of the type used in Chapter 4 with two events A and B specially marked. Suppose we have a body which includes these two events along its world-line. The

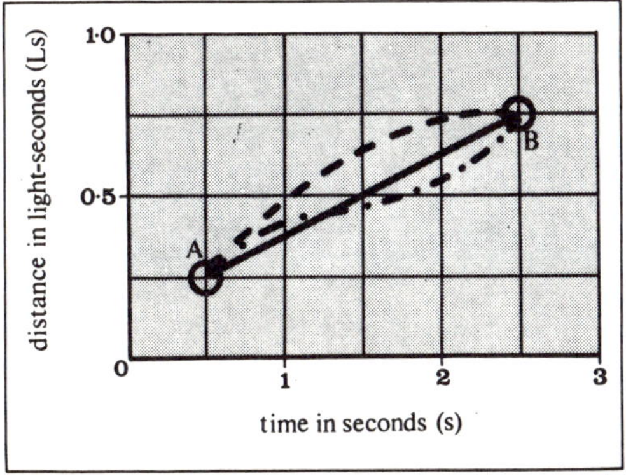

Fig. 7.1. Space-time diagram showing two events joined by three possible world-lines. The straight world-line (shown as a solid line) is the only world-line joining A and B which represents natural unforced motion.

world-line of such a body could be any line on the diagram joining A and B, provided that the slope of the line was always less than 1 Ls/s, and three possible world-lines for the body are shown. The only one of all the possible world-lines which corresponds to the natural motion of the body is the straight line shown as a solid line on the diagram. All the other lines are curved and correspond to forced motions involving accelerations. So in the case of special relativistic space-time and the old definition of natural motion (Newton's first law), we find that a special kind of world-line, a straight line, represents a naturally moving body (see the beginning of Chapter 5).

As the second stage in our digression let us go back a little further and consider again a flat surface like that represented in Fig. 1.6. The positions of points on such a surface can, of course, be represented by a pair of cartesian space coordinates and so it is called a two-dimensional surface. We can join any two points on such a surface (like points A and B in Fig. 1.6) by any number of lines, three of which are shown on the diagram. If we now ask which of all these lines is the shortest, then we know it is the straight line shown as a solid line on the diagram. Also we know we can find the distance along this straight line between two points such as A and B in terms of the coordinates of A and B by using the method attributed to Pythagoras, which we discussed in Chapter 3.

Now as the third stage in our digression let us consider another kind of two-dimensional surface, the surface of a sphere. Such a surface can also be considered to be two-dimensional since the position of any point on the surface can be specified by two coordinates such as the coordinates of latitude and longitude used to specify the positions of points on the earth's surface (see Fig. 7.2(a)). Such coordinates are not cartesian but are perfectly satisfactory for specifying the position of any point on such a surface. If we consider any two points A and B on this surface then we can, once again, join them by drawing any number of lines on the surface, three of which are shown in Fig. 7.2(b). If we now ask which of these lines is the shortest, then the answer is not so obvious as it was on the flat surface. However, it is obviously a key question in maritime navigation to ask what is the shortest line or route between any two points on the earth's surface and the answer has long been known. It is that the shortest line is the line on the earth's surface obtained if we imagine cutting the earth exactly in half so that both the points A and

B lie on the cut (see Fig. 7.2(c)). Such a line is called a great circle and it is the exact analogy on a spherical surface to a straight line on a flat surface.

The distance along the great circle joining any two points like A and B can be found in terms of the coordinates of A and B by means of a calculation which is something like the Pythagoras calculation for a flat surface. However, the calculation for a spherical surface differs in two significant ways from the Pythagoras calculation. First, it is most easily carried out by splitting the great circle into a large number of very short segments, calculating the length of each segment in terms of the coordinates of the points at each end of the segment, and then adding up the lengths of all the segments to give the total great circle length between A and B. Secondly, the calculation of the length of each segment, while it does give the square of this length as the sum of terms obtained by multiplying coordinates

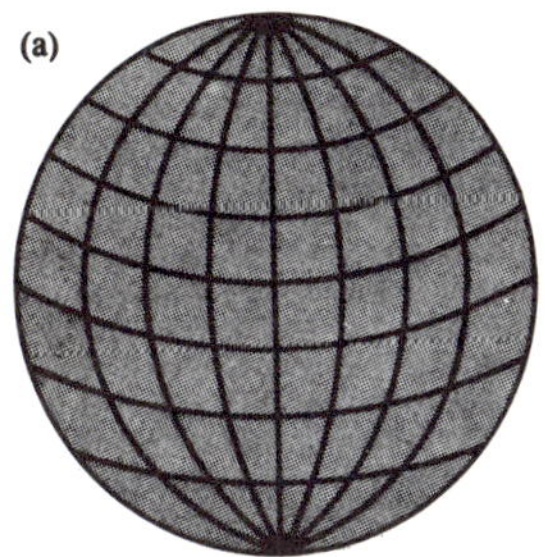

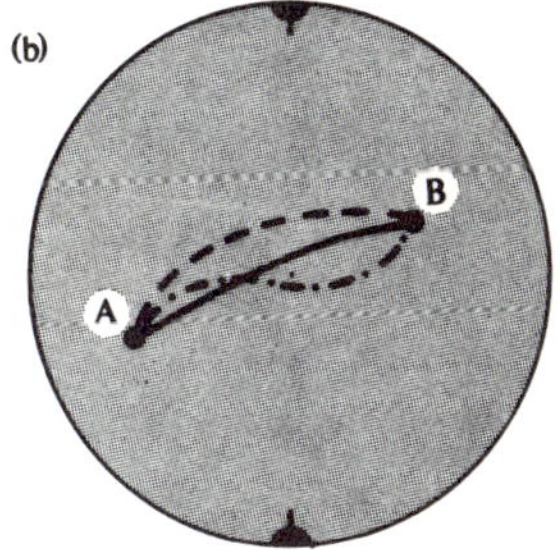

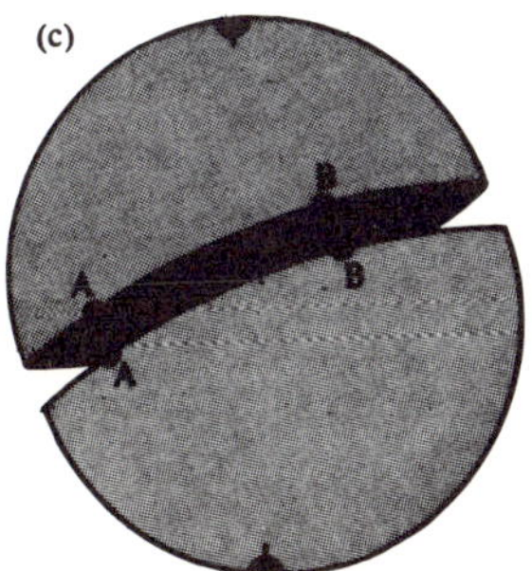

Fig. 7.2. Surface of a sphere showing (a) lines of latitude and longitude, (b) two points A and B joined by three possible lines (the solid line is the great circle or geodesic joining A and B) and (c) a method for finding the great circle joining any two points on the surface.

together, is of a different mathematical type from the Pythagoras calculation. It is difficult, however, to explain just how it is different without discussing in detail the work of the famous mathematician Gauss on the geometries of all kinds of two-dimensional surface.

Gauss showed that special kinds of line, like straight lines on flat surfaces and great circles on spherical surfaces, exist for any kind of two-dimensional surface; such lines are called *geodesic* lines or *geodesics*. Methods of calculating distances along geodesics, similar to those used for straight lines and great circles, also always exist and Gauss was able to show that subtle differences in these calculations could form the basis of a method of classifying all possible two-dimensional surfaces into different types. The complexity of the mathematics involved means that we cannot discuss this classification in any more detail here. However, some idea of the pitfalls of such an attempt at classification can be obtained if we notice that Gauss classifies a flat surface and the surface of a cylinder together as being of the same type, while a spherical surface is of a different type. One clue to help us to accept this perhaps unexpected result is that a flat sheet of paper can be neatly wrapped round a cylinder but not round a sphere.

Now let us see how this long digression is relevant to our discussion of general relativity. We can best understand what Einstein did if we make a series of analogies or comparisons between the various stages of the digression. First compare the simplified two-dimensional space-time of special relativity (which we used in the space-time diagrams of Chapter 4) with a two-dimensional flat surface. We can draw an analogy between the straight line in space-time representing the natural motion of a body and the straight line or geodesic on the flat surface representing the shortest distance between two points on the surface. We might press the analogy between the space-time of special relativity and a flat surface and hence refer to this kind of space-time as a 'flat' space-time.

Next compare the two-dimensional flat surface with the other types of two-dimensional surfaces studied by Gauss and draw the analogy, which is very close in this case, between the geodesics on the flat surface (straight lines) and the geodesics on the other kinds of surface (one example being the great circles on spherical surfaces). These other kinds of surface are called curved surfaces of different types of curvature (although it must be remembered that this description is not completely straightforward, since the surface of a cylinder is classified as a 'flat' surface).

Next let us speculate that since many types of two-dimensional curved surface exist which are related to a two-dimensional flat surface, why should there not be the equivalent of different types of curved surface in two-dimensional space-time complete with the equivalent of geodesics? In fact mathematical equations can be written which represent such types of surface and their geodesics, and which are exactly analogous to the mathematical equations of Gauss' work.

Finally, let us complete this series of analogies by making a comparison between the 'flat' space-time of special relativity, with its straight-line geodesics, and these curved space-times of different types, also with their geodesics. And let us now make the enormous guess, or leap in the dark, of assuming that the geodesics in the curved space-times correspond to the natural motions of Einstein's first law of motion in just the same way as the geodesics in flat space-time correspond to the natural motions of Newton's first law of motion. Since Einstein's first law includes both inertia and gravitation, we have to add to our guess the assumption that different gravitational environments would correspond to curved space-times of different types with correspondingly different geodesics.

These enormous guesses are very similar to the sort of intuitive assumptions Einstein made in working out the details of general relativity. They are not precisely similar, since we have not mentioned the generalization of the work of Gauss to spaces of three or more dimensions which was carried out by Riemann and others in the nineteenth century. Nor have we talked in terms of four-dimensional space-time as Einstein did. Our analogies, however, retain nearly all the main points of Einstein's work. The striking thing about this final step of identifying the natural motions of bodies with geodesics in curved space-times is how improbable and incredible it seems, and yet extraordinarily enough it does work. Of course in a book of this type we cannot deal with the detailed working out—the mathematics involved, the general theory of tensors, is unfamiliar even to most graduate physicists—but we can discuss the main results in outline and so get some idea of what the theory can do.

The first and easiest application of general relativity is to a region which does not contain any matter at all and which is so far from any matter that gravitational effects can be considered to be negligible. In such an empty region, general relativity predicts a flat space-time with straight-line geodesics; in other words, it predicts the space time of special relativity and thus shows the correct asymptotic behaviour.

The next application of general relativity was made to a region of empty space-time which surrounds a spherically symmetrical non-empty region; this approximates to the situation in the solar system in which we have the spherically symmetrical massive sun surrounded by what is largely empty space. In this case the equations of general relativity predict a curved space-time with geodesics which can be calculated. And when they are calculated, these geodesics turn out to correspond to motions which are almost identical with the planetary orbits as calculated by classical physics using Newton's gravitational law (see Chapter 3). This was a great success for the theory, since it means that it is equivalent or asymptotic to Newton's theory in this case and so all the evidence for Newton's theory of planetary motion is also evidence for general relativity. However, this was not all. The geodesics corresponded to motions which were almost but not quite identical with the classical orbits. The differences were so small that for all the planets except one they were less than the accuracy with which the orbits had been measured. The exception was the orbit of the planet Mercury, for which there existed a sizeable discrepancy between the measured orbit and the orbit predicted by Newton's theory (see Chapter 3). And this discrepancy is almost exactly equal to the difference between the orbits calculated by general relativity and by Newton's theory. So general relativity is not only asymptotic to the classical theory; it has cleared up an anomaly which the classical theory was unable to explain.

The third application of general relativity is to the motion of light in a region of space-time like that considered above in the case of planetary motion. Einstein assumed that a light signal would travel along a geodesic in space-time for which the interval between any two adjacent events along the geodesic is zero. Such geodesics are called null geodesics and are generalizations of the straight-line geodesics corresponding to the paths of light rays in special relativity; these straight-line geodesics also had the property that the interval between any two events along them was zero (see Chapter 4).

Calculations of these null geodesics for light signals from distant stars which just grazed the surface of the sun on their way to the earth led to the prediction that such light-signals should be deflected from straight paths by very small amounts. Unfortunately these deflections cannot be measured very accurately, since they are so small. However, the measurements that are available are in fairly good agreement with the predictions of general relativity and certainly agree better with these predictions than they do with the predicted deflections derived from classical physics.

The fourth application is to a region in which the effect of gravitation is constant. This is like the situation in any laboratory on the earth's surface in which the acceleration of any freely falling body is the same in both magnitude and direction at all points in the laboratory. This acceleration is not the same over the whole of the earth's surface, since its direction at any point is always approximately perpendicular to the earth's surface at that point and so the direction of the acceleration will be different at every point due to the curvature of the earth's surface. Also the magnitude of the acceleration does differ slightly from point to point. However, for any region near the earth's surface whose dimensions are small compared with the earth's radius, this acceleration can be considered to be constant for all practical purposes. In such a region general relativity again leads to a flat space-time and the equations of special relativity are applicable.

And that unfortunately is almost the full extent of the contact between general relativity and experimental observations. Another experimental effect, the influence of gravitation on the time-keeping properties of atomic clocks, does in fact occur, although it is now generally agreed that this effect is not a consequence of Einstein's detailed general theory, but can be predicted from the principle of equivalence. There are some other situations in which general relativity does predict substantially different effects from special or classical relativity, but these situations involve either extremely massive bodies or extremely large regions of space. Hence they are not capable of being tested on earth unless nature is so good as to provide such situations at positions in the universe which are near enough to us for earthbound astronomers to observe them accurately and in sufficient detail. Even the two experimental tests of general relativity described above were only possible because nature had been kind enough to provide a suitable massive object, the sun, at not too great a distance from us. Purely terrestrial experiments to test general relativity may one day become possible, but they seem at the moment to be well beyond the limits of experimental technique.

The theory of general relativity, then, succeeds in its object of uniting gravitation and inertia into one consistent theory. And it is a theory which has the right asymptotic behaviour in two ways. First, it reduces to the classical gravitational theory under certain conditions, and secondly it reduces to special relativity when gravitational effects are negligible or constant. Therefore in one sense all the evidence for classical dynamics and gravitational theory, and for special relativity can also be said to support general relativity. Further than that, wherever there are observable differences between the predictions of general relativity and previous theories, the evidence is in favour of general relativity. However, as we have seen, such differences are few and the weight of experimental evidence for general relativity is far less than that for special relativity.

The second main object of general relativity, that of removing the restriction to inertial observers, is also achieved, since it turns out that the general theory of tensors can employ any system of co-ordinates pertaining to any kind of observer in its treatment of world-lines, events and intervals.

The general theory of relativity is therefore very different from the special theory. It succeeds in working out in detail a new basic law of motion which combines inertia and gravitation in such a way that they are both presented as part of the very structure of space and time. In effect it says that a body moves in the way it does under inertia and gravitation because it can do nothing else. However, the theory does not deal explicitly with electromagnetic or nuclear forces, except in the two simple cases mentioned above where it reduces to special relativity. We could say that Einstein has triumphantly rewritten Newton's first law of motion but has not given a general relativistic replacement for Newton's second law of motion.

8
Relativity in Perspective

We have now given some account of the theories of relativity, of how they arose and how they are related to classical ideas and to one another. In a sense then the main object of this book has been attained. In this final chapter we shall briefly review the present and possible future status of the theories and touch on one or two related and interesting topics.

The special theory of relativity has had, and still has, its share of fierce critics and detractors. However, it is supported by such a vast amount of experimental evidence (see Chapters 4, 5 and 6) that it is very difficult seriously to doubt it. The body of evidence is so wide-ranging that it resembles the evidence which had been amassed in support of classical dynamics by the beginning of this century. Therefore if any theory should ever be devised which supersedes special relativity, it must surely be set up in such a way that special relativity will remain as a special case of the new theory. To put it another way, any new theory must be asymptotic to special relativity just as special relativity is asymptotic to classical relativity and classical dynamics. The principles and details of special relativity are probably best regarded not as absolutely true but as the best and most concise way yet devised for describing a vast amount of experimental data.

The general theory of relativity, on the other hand, is much less secure. There are only a few experimental tests of the theory and these all involve measurements of extreme difficulty which are at or near the limits of the accuracy at present attainable. The best experimental test, the discrepancy in the orbit of Mercury, is not totally convincing, since the observed discrepancy is in fact a residual discrepancy after several substantial corrections have been made to the initial observations. The effect actually observed is more than a hundred times greater than the discrepancy for which general relativity accounts. Also it has recently been suggested that the discrepancy could be explained if the sun should turn out to be not quite spherical in shape. The degree of flattening required to make a substantial contribution to the discrepancy is only about one part in ten thousand and careful experiments are now going on to see if such a flattening can be directly measured by optical observation of the sun.

These doubts, however, are mainly centred on the detailed working out of the theory of general relativity. They do not mean that the basic principle, the principle of the equivalence of gravitation and inertia, is necessarily suspect. And in fact the evidence for this principle (see Chapter 7) is generally considered to be extremely strong. A recent development in this field has been the working out of a whole series of theories based on the same basic principles as general relativity but differing in the detailed mathematical way in which the principles are incorporated into the theories. From the point of view of this work, Einstein's general relativity is seen as just one theory out of a whole class of theories all stemming from the principle of equivalence. General relativity is the simplest of these theories, but of course simplicity alone is no argument in its favour.

Another development, first started by Einstein and since carried on by a large number of other theoretical physicists, has been an attempt to incorporate electromagnetic and nuclear forces into a unified theory in a way similar to that in which gravitation was incorporated into general relativity. Despite much effort, however, this work does not seem to have reached a satisfactory conclusion.

Another aspect of gravitation, which has also been the subject of much study, is the possibility of observing gravitational waves produced by the rapid movement of massive objects in much the same way as electromagnetic waves are produced by the rapid movement of electric charges. The theory of gravitational waves is so much more complicated than that of electromagnetic waves and any possible experiments to detect them are so difficult that progress has been slow. Recently, however, in experiments involving the use of sensitive seismometers, there have been reports of the detection of some very small effects which may be attributable to gravitational waves reaching the earth from somewhere in the universe.

Finally let us briefly mention a topic which is basic to the whole subject of dynamics and that is to ask what is the source of the inertia of matter, or, alternatively, why it is that inertial motion is distinguished from all other types of motion and inertial observers are in a class by themselves. This subject was considered as long ago as 1880 by Mach in a book on the history of dynamics in which he suggested that the inertial properties of any particular piece of matter were in some way attributable to the influence of all the other matter in the universe. Unfortunately he did not give any detailed theory of this influence but the basic idea, or Mach's principle as it is called, has intrigued many people ever since. The principle seems to imply that if, in some way, say half the matter in the universe were to be removed, then the inertial properties of the remaining matter would be altered and gravitational forces would be considerably reduced. Unfortunately for a scientific test of Mach's principle, this sort of experiment is difficult to perform! However, efforts are being made to try and work out a detailed theory of the interaction involved so that some more easily testable predictions can be madc.

The resolution of all these tentative ideas and suggestions most probably depends on a combination of three things, the improvement of experimental techniques in terrestrial measurements, the discovery of suitable events to study in that natural laboratory, the universe, and finally the further development of the science of cosmology, the science which studies the nature, origin and history of the universe as a whole. All these things are difficult and so progress is likely to be slow. It is also likely, if past experience is any guide, that the resolution of one problem or difficulty will bring into view another more subtle and elusive problem for the next generation of scientists to tackle.

At about this point in books on relativity it is almost traditional for something to be said on the relation between relativity and philosophy. This seems to me to be the sort of enormous extrapolation that we learnt to suspect in Chapter 1 and so I shall forgo the natural temptation of an author to expose his readers to his own personal philosophical views. I shall content myself merely with saying that we might absorb two things from this account of the theories of relativity—first, that traditional ideas are not necessarily the whole truth about anything, no matter for how long they have been held or how many people may subscribe to them, and secondly, that the beauty, simplicity or intuitive appeal of any concept is no sure guide to its validity, since the world or the universe may well be ugly, complicated and inaccessible to our intuition.

The main aim of this book has been to present the basic ideas of relativity in a way that would be comprehensible to people without any great mathematical or scientific knowledge. I hope that this has been achieved and that the book will also be useful to those with some command of mathematics, since it provides much more verbal explanation than is usual in books on relativity. As a final warning let me stress that this kind of verbal exposition is only part and probably the lesser part of any scientific theory. It is incomplete without a detailed mathematical exposition of the theory and a quantitative comparison between theoretical predictions and experimental measurements. All the plausible verbal arguments that Einstein could muster to support his ideas would have been long forgotten if he had been unable to express these ideas quantitatively and if precise experiments had not at least partly confirmed his detailed calculations.

Appendix 1
Standards, Units and Abbreviations

Unit of length

The internationally accepted length unit is the metre (m) which is defined as 1650763·73 times the wavelength in vacuum of the orange-red light emitted by the isotope of mass number 86 of the gaseous element krypton when it is excited in an electrical discharge.

Unit of time

The internationally accepted time unit is the second (s) which is defined as 9192631770 times the period of vibration of a particular type of radar radiation emitted by the isotope of mass number 133 of the element caesium when it is excited in a suitable way.

The speed of light and other electromagnetic radiations

Many types of electromagnetic radiation are known to exist. They include radio and radar signals, infra-red radiations, visible light, ultraviolet radiations, x-rays and γ-rays. All these radiations have been found to travel at the same speed in empty space. The best measurement available so far of this speed is $2{\cdot}99793 \times 10^8$ m/s. In this book we have very often made a slight approximation and taken this speed to be $3{\cdot}00 \times 10^8$ m/s.

Abbreviations

The main units and abbreviations used in this book are shown in the following table:

Physical Quantity	Unit	Abbreviation
Length	metre	m
	kilometre (1000 m)	km
	light-second (3×10^8 m)	Ls
Time	second	s
	hour	h
	year	yr
Velocity or speed	kilometres per hour	km/h
	light-seconds per second	Ls/s
	metres per second	m/s
Acceleration	(metres per second) per second	m/s^2
Mass	kilogramme	kg
	'unusual' unit of Chapter 5	($3{\cdot}33 \times 10^{-9}$) kg
Force	newton	N
Momentum	kilogramme metres per second	kg m/s
Energy	joule	J
	'unusual' unit of Chapter 5	(3×10^8) J

Approximate conversion factors for British and U.S. units of mass and length

1 pound (lb)=0·4536 kg
1 kilogramme (kg)=2·205 lb

1 yard (yd)=0·9144 m
1 metre (m)=1·094 yd

Appendix 2
Suggestions for further Reading

Books which discuss relativity using very little mathematics

Bondi, H., *Assumption and Myth in Physical Theory*, Cambridge University Press, 1967.

Russell, B., *The ABC of Relativity*, George Allen & Unwin, 1969.

Books which give simple accounts of relativity using some mathematics

Eddington, A., *Space, Time and Gravitation*, Cambridge University Press, 1966.

Rosser, W. G. V., *Relativity and High Energy Physics*, Wykeham Publications, 1969.

Books which give detailed mathematical treatments of relativity

Bondi, H., *Relativity* (Reports on Progress in Physics, vol. 22), The Physical Society, London, 1959.

Dicke, R. H., *The Theoretical Significance of Experimental Relativity*, Blackie, 1964.

Einstein, A., and others, *The Principle of Relativity*, Dover Publications, 1923 (this book contains several of Einstein's original papers).

Glossary

All the key technical terms used in this book are carefully defined when they are first used in the main text. The reader is therefore recommended to refer to these definitions with the help of the index. However, the following list of brief definitions may also be helpful.

Physics is the science which deals with matter and energy and their interactions in the fields of mechanics, dynamics, acoustics, optics, heat, electricity, magnetism, radiation, and atomic and nuclear phenomena.

Observers are the individual scientists who make the measurements which form the basis of the science of physics.

Relativity theories deal with how the laws of physics appear to two or more observers who are moving with respect to one another.

Classical physics denotes the physical laws and theories which had been established at the end of the nineteenth century.

Classical relativity is the term used for the relativity theory which had been accepted up to the end of the nineteenth century.

Inertial observers are observers who are moving with respect to one another at constant velocities.

Non-inertial observers are observers who are moving with respect to one another at varying velocities.

Special relativity is the new relativity theory, introduced by Einstein in 1905, which deals with how the laws of physics appear to two or more inertial observers.

General relativity is the new relativity theory, introduced by Einstein in 1916, which deals with how the laws of physics appear to two or more non-inertial observers and which incorporates gravitation as an integral part of the theory.

Index